Liberalization and Entrepreneurship

Liberalization and Entrepreneurship

Dynamics of Reform in Socialism and Capitalism

BRANKO MILANOVIĆ

M. E. SHARPE, INC.
Armonk, New York
London, England

Available in the United Kingdom and Europe from M. E. Sharpe,
Publishers, 3 Henrietta Street, London WC2E 8LU.

Library of Congress Cataloging-in-Publication Data

Milanović, Branko.
    Liberalization and entrepreneurship : dynamics of reform in
socialism and capitalism / by Branko Milanovic.
        p.   cm.
    Bibliography: p.
    Includes index.
    ISBN 0-87332-568-0
    1. Economic policy. 2. Deregulation. 3. Comparative economics.
4. Socialism. 5. Capitalism. I. Title.
HD75.M56   1989   89-4196
338.9--dc19   CIP
                                                       rev.

Printed in the United States of America

BB 10 9 8 7 6 5 4 3 2 1

CONTENTS

Figures   vii
Tables   ix
Acknowledgments   xi
Introduction   xiii

Part I. THE MEANING OF LIBERALIZATION

**1. Modes of Organization of Production
and the Legal Framework**   3

1.1 Capital, Labor, and Entrepreneurship   3
1.2 Classification of the Modes of Production   4
1.3 Labor as a Hired Input   6
1.4 Workers as Entrepreneurs   10
1.5 The Distribution of Income   11
1.6 Economic Coordination   11
1.7 Firms' Maximands   13
1.8 Capitalist Economies   14
1.9 Socialist Economies   19
1.10 The Structure of Production and the Legal Framework   22

2. The Economic Role of the State   27

2.1 Definition of State Interference   27
2.2 Mode of Production and State Intervention   29
2.3 State Intervention in Capitalism and Socialism   31
2.4 What Is Liberalization?   34

**3. The Efficiency of Different Modes
of Production**   37

4. Ideological Grounds for Liberalization   42

4.1 Manifestation of a Changing Attitude toward
        the Role of the State   42
4.2 Central Planning and Development   47
4.3 The Failure of State Socialism   49

Part II. THE POLITICS OF LIBERALIZATION

5. Reform in State Socialism   55

5.1   The Economic Structure of State Socialism   55
5.2   The Social Composition   57
5.3   Liberalization: Reduction of the State Sector   59
5.4   Liberalization: Expansion of the Private Sector   65
5.5   The Central Conflict in Socialism   74
5.6   The Politics of Reform   79
5.7   The Calculus of Reform   85
5.8   Applications: Poland, China, the Soviet Union   88

6. Reform in Developed Capitalism   96

6.1   The Economic Structure of Developed Capitalism   96
6.2   The Social Composition   99
        Excursus: Political Elites and Freedom of Action   101
6.3   What Is Liberalization?   103
6.4   Deregulation   105
6.5   Reduction of Subsidies and Lowering of Protection   108
6.6   Scaling Down the State Sector   109
6.7   Capitalists and Privatization   113
6.8   The Position of State-Sector Managers and Workers   115
6.9   What Future for the State Sector?   117
6.10 The Central Conflict in Capitalism   119
6.11 Would Elimination of the Central Conflict in Capitalism
        Lead to the Disappearance of the Capitalist Mode of
        Production?   124

Part III. A POSSIBLE OUTCOME

7. The Age of Technocrats?   127

7.1   Introduction   127
7.2   The Ascent of the Problem-Solvers   128
7.3   Unification of Elites   133
7.4   Obstacles to Unification   137
7.5   Prospects for Peace   146

Annex 1: Sources of Data   149
Notes   153
References   173
Name Index   177
Subject Index   179
About the Author   183

FIGURES

4.1   Center-right strength in four major countries   43
5.1   Social structure in state socialism   60
5.2   Poland: share of non-state sector in national income   74
6.1   Social structure in capitalism   102

TABLES

1.1	Alternative ways of organizing production and ownership	5
1.2	Distribution of the entrepreneurial role	10
1.3	Recipients of income by mode of production and type of income	12
1.4	Importance of public corporations and state sector in some OECD countries	15
1.5	Importance of public and state sector in some LDCs	17
1.6	Importance of cooperative-1 type in selected capitalist countries	19
1.7	Importance of the state sector in some socialist countries	20
1.8	Relative importance of different modes of production	22
1.9	Alternative modes of organization of production in Poland and the United Kingdom	23
	Annex to Table 1.9	152
1.10	The structure of production in the United Kingdom, 1971 and 1981	25
3.1	Reasons for inefficiency	40
4.1	Japan's and Romania's rates of GDP growth (in notes to chapter)	159
4.2	Comparison of Czechoslovakia and Hungary with Austria and Greece (in notes to chapter)	160
5.1	Appropriation of income in state socialism	57
5.2	Poland: percentage of centrally allocated inputs	63
5.3	Liberalization in state socialism	64
5.4	Reform in state socialism	72
5.5	China and Hungary: the structure of production	73
5.6	Czechoslovakia: the structure of production (in notes to chapter)	163

5.7	Tally of anti- and pro-reform social forces	85
5.8	The calculus of reform I	86
5.9	The calculus of reform II	87
6.1	Appropriation of income in developed capitalism	100
6.2	Origin of income of different social groups	104
6.3	Offer and opening prices	112
6.4	Estimated number of shareholders (in notes to chapter)	166
6.5	Rankings by different social groups of alternative organization of the state sector	118

ACKNOWLEDGMENTS

This is a book about reforms in capitalist and socialist economies and the world that may emerge if they are successful. The idea of considering the reforms as two species of the same fundamental movement away from *dirigiste* economic policies emerged several years ago. Conversations that I had with Mr. Slaheddine Khenissi from the World Bank—even if he may not recognize many of our topics of discussion, much less agree with a number of my views—were instrumental in helping me to realize the link between the two types of reforms and in alerting me to the need to devise a unified framework within which to study them.

Parts of the book were read by Ms. Michele de Nevers, Mr. Srdja Trifkovic, and Mr. Shangzhi Wu. They contributed valuable comments, particularly to chapter 5, which deals with the political economy of socialism. I hope to return more fully to the topic of reform and socialist economic system in another volume.

I would also like to thank Ms. Patricia Kolb, executive editor at M. E. Sharpe, who read the manuscript with great care and improved both its language and its substance.

In view of the recent widespread misuse of the word "liberal," it may be useful to state at the beginning that it is used in its classical, European, not its contemporary American sense.

The views expressed in the book are my own. They should not be attributed to any of the institutions with which I am or have been affiliated.

INTRODUCTION

This book starts from the premise that economic liberalization—that is, reduced state interference in economic life—is the common element in the current trend toward privatization and deregulation in the West and the reform of economic policies and institutions in the East. In popular parlance, "privatization" and "*perestroika*" are its watchwords, Margaret Thatcher and Mikhail Gorbachev its heralds. But the moves toward liberalization, and the nature of liberalization, will not be the same everywhere. They will be determined by the social characteristics of different societies and by the correlation of forces between different social (anti- and pro-reform) groups.

Although these movements are widely recognized and discussed, they are usually examined in an *ad hoc* fashion, and the link between the reforms in the West and those in the East is not well understood. Both reform movements spring from the same intellectual source: dissatisfaction with *dirigiste* policies. They respond to the same realization: that at the current stage of development, technological progress requires decentralization of initiative and risk-taking. In order to study the reform process in the two systems, to assess similarities and differences in the way it may proceed, it is necessary to dispose of a general conceptual framework capable of embracing both a (predominantly) market economy and a (predominantly) centrally planned economy. The key objective of this work is to provide such a unified framework. Then, starting from this common conceptual basis, we can address the problems associated with liberalization in the two systems, the social groups likely to benefit or lose in it, and the obstacles to its implementation, as well as further prospects.

The organization of the book is as follows.

The first two chapters set out a general conceptual framework within which the economic and social structure of a society can be studied. We use this framework to capture the essential differences between capitalist and socialist economies. We then define exactly what state intervention in the economy (and its converse: liberalization) mean. In the following two chapters we consider advantages and disadvantages of different ways the process of production can be organized (i.e., different ways in which labor, capital, and entrepreneurship can be combined) and the ideological factors that underlie the current shift toward less state interference. The 1980s are, it would seem, one of those periods of ideological ferment when old dogmas are questioned and new solutions proposed. But, unlike, say, in the 1950s, it is an accent on economic freedom and a general skepticism about the role of the state that prevail.

The second part of the book (chapters 5 and 6) takes the discussion to the realm of applications. It analyzes reforms recently undertaken in a number of countries, and the relationship between them and a strictly defined liberalization. The question is to determine whether these policies, often associated with liberalization, do indeed coincide with it or represent a covert way in which special interests, paying lip service to the ideological *esprit de temps*, camouflage themselves to better pursue their own objectives. We are thus led to the domain of the political economy of liberalization in state socialism and capitalism. We shall try to find out what kinds of reforms can realistically be expected in the two systems.

Will the reforms make the two systems more similar, or will important differences remain? The last chapter attempts to provide an answer to this question. It sets out to investigate how a world in which the economic sphere was more emancipated from state tutelage might look; what changes in international economic and political relations might be expected if societies both more technocratic and more liberal (in the economic sense) emerged both in the West and, particularly, in the East. What would the prospects for economic cooperation and peace then be? To put it differently, would the diminished role of the state and greater economic freedom lead to a world in which one of the main causes of conflict (nationalism, supported and expressed by the state) would be eliminated, or, on the contrary, would a supposed or real liberalization favor specific social groups which would become able to harness the power of the state to further their particular interests?

Part I

THE MEANING OF LIBERALIZATION

Modes of Organization of Production and the Legal Framework

"[A] certain mode of production, or industrial stage, is always combined with a certain mode of cooperation, or social stage, and this mode of cooperation is itself a 'productive force.'"—Marx*

1.1 Capital, Labor, and Entrepreneurship

We begin by considering the different ways in which the process of production may be organized—that is, the modalities of ownership over the instruments of production and labor and the exercise of the entrepreneurial function. Labor—or, more exactly, workers—can, as historical experience shows, be deprived of the right of ownership over their own persons, or be legally free. Not having the right of ownership over oneself implies absence of the right to choose one's own occupation, conditions of work, place of residence, etc. This has been the case in slavery and in feudal systems. By contrast, when labor is legally free it is not legally possible for one person to own another person. Legally free labor can be hired labor. When this is the case, the organization of the process of production is (generally) carried out by the owners of capital. Alternatively, legally free workers may themselves organize production.

---

*Karl Marx, *The German Ideology*, in Robert C. Tucker (1978, p. 157).

Definition 1. The factor that organizes the process of production is the one that is the claimant of the residual income, i.e., of the income that remains after payments for all the hired factors of production. It is obvious that, theoretically, the role of the organizer of production, i.e., the *entrepreneurial role*, can belong to the owners of one or another factor of production (i.e., to either workers or capitalists) or to a specific factor of production whose role consists in combining capital and labor.

Capital can be either privately owned or state-owned. On the basis of ownership, capitalists (whether private individuals or the state) may organize the process of production. Alternatively, capital may be leased to either workers or entrepreneurs who organize production.

Definition 2. The *economic owner* of capital or any other asset is the agent (private or state) who receives income from that asset. The agent who, from the strictly legal point of view, owns an asset need not always coincide with the agent who receives the stream of income from that asset. (For example, a tenant who lives in a privately owned apartment that is subject to rent control implicitly receives a portion of the return from the capital. Or, workers who work on freely supplied state capital actually appropriate capital income, even if the state remains a nominal owner.)

Table 1.1 shows alternative ways in which (a) ownership and (b) the entrepreneurial role in the process of production can be distributed.

Definition 3. A particular single combination of ownership and entrepreneurship will be called a type or *mode of production*.

1.2 Classification of the Modes of Production

Let us examine each of the modes of production more closely. Consider the principle around which Table 1.1 is organized. The outside titles show that, in the most general terms, capital can be either privately owned or state-owned, while labor can be legally unfree or legally free. These are the two broadest classifications. When capital is privately owned, its owners can either exercise the entrepreneurial role or hire out their capital. Intuitively, this is the most obvious difference between capitalist-entrepreneurs and rentiers. The same distinction between the

Table 1.1

Alternative Ways of Organizing Production and Ownership

CAPITAL \ LABOR	Without right of ownership (1)	With right of ownership (2) Has entrepreneurial role	(3) Is hired out	The system coordination (4)
PRIVATE (1) Has entrepreneurial role	Slavery Feudalism	Owner-workers (self-employed Cooperative-1)	Capitalism	Decentralized
PRIVATE (2) Is hired out	n.e.	Cooperative-2	Entrepreneurial system	Decentralized
STATE (3) Has entrepreneurial role	Asiatic mode of production; Mayas; Gulag	n.e.	State socialism	Centralized
STATE (4) Is hired out	Chinese communes	Labor-managed	Public corporation	Decentralized

n.e. = nonexistent.

entrepreneurial role and hiring out of the factor exists when capital is state-owned. It also exists with respect to legally free labor: it can take over the entrepreneurial function or be hired out. We thus have entirely symmetrical situations for capital and labor. The only exception is the case of labor that is not legally free and, therefore, cannot be hired (for this presupposes a free contract), nor can it normally, for obvious reasons, take over the entrepreneurial function.

We shall consider column 1 in Table 1.1 first.

When there is no prohibition on ownership of labor (in other words, no legally enforced freedom of labor), labor and capital appear clearly as merely two factors of production. The legal position of a worker or of a machine is exactly the same. This identity is best reflected in Aristotle's famous comment about slaves being animate tools. The legal difference between labor and capital—which we now take for granted—is not even noticed.[1] In slavery and feudalism, the owner of workers and machines also performs the entrepreneurial function. We thus find slavery and feudalism located in cell (1,1).[2]

The only other systems in which workers do not have the right of ownership over their own persons belong to the species that Marx called the ''Asiatic mode of production'' (also known as ''hydraulic systems"). These are systems in which the state owns the capital, commands the legally unfree labor, and organizes the process of production. Apart from Asia, the system existed among the Mayas, the Incas, and in Mycenae, while in the modern era it appears only incidentally: in the Soviet Union's "Gulag" system, and to some extent in China's communes from the mid–1950s to the late 1970s and Soviet kolkhozes (up to the mid–1950s).[3] Communes were a rather paradoxical case because the forced nature of their formation and the absence of freedom of laborers to leave the communes put them in a class of systems with bonded labor. On the other hand, an entrepreneurial role was clearly exercised by these same laborers, whose incomes depended on the results of the commune. To be quite accurate we therefore need to divide column (1) into two parts: unfree labor with an entrepreneurial role (communes and kolkhozes) and without one (all other systems with unfree labor).

1.3 Labor as a Hired Input

In modern economic systems legal freedom of labor is guaranteed. Therefore, in the rest of our analysis we shall consider only systems whose legal framework (see Definition 6, below) bans the existence of unfree labor. We are thus concerned only with columns 2 and 3 in Table 1.1.

We will first look at column 3. When private ownership of capital is combined with hired labor (cell (1,3)) we have *capitalism*. The entrepreneurial role generally belongs to owners of capital. Under capitalism we can distinguish between two types of organization: the *joint-stock company*, whose shares are publicly traded, their holding practically open to all individuals; and the *limited liability company*, whose shareholding is limited and nonpublic. In the first, management is partially dissociated from ownership; the entrepreneurial function is exercised jointly by the owners of capital (capitalists) and the management of the firm. In the second type, the entrepreneurial role belongs only to the owners of capital.

The recent spate of management buy-outs represents a movement from the joint-stock company back toward the classical capitalist firm with coincidental ownership and management rights. In a sense it represents a full circle in capitalist development. In the original capitalist firms, management rights were derived from property rights; joint-stock companies diluted the entrepreneurial role of capitalists; finally, management buy-outs again unify the two roles, but now it is entrepreneurship that serves as a stepping stone for acquisition of property.

If both privately owned capital and labor are hired, we have an *entrepreneurial mode of production*. This is the type that is normally assumed in neoclassical economic theory. A specific factor of production, embodied in the entrepreneur, hires both (privately owned) capital and (legally free) labor. This is shown in cell (2,3). According to Fernand Braudel (1984, pp. 129–30), the distinction between capitalist and entrepreneur was already obvious in Venice in the eleventh century.[4] A commercial loan (pure capitalist loan) was often associated with a partnership, or *colleganza*, agreement. Under the agreement one party advanced a sum of money to a traveling partner, and after the voyage (usually a sea voyage) and the repayment of the originally borrowed amount, the traveler kept one-quarter of the profit and paid three-quarters to the capitalist. We see that in this case the rate of interest (purely capitalist income) is not fixed, say, at so many percent per year, but is determined in terms of a (variable) profit. The key feature of capitalist income, however, is not that it is a fixed income, but that it is an income resulting from ownership of assets, and not from a direct role in the process of production. In the example, the return on pure entrepreneurship was one-quarter of the realized profit.

Hired labor combined with state ownership over capital is *state socialism* (cell (3,3)). In this system the state is also the entrepreneur: it decides about the organization of production. Thus, state socialism differs from capitalism only in ownership of capital.

When state ownership of capital is combined with hired labor and the entrepreneurial role is transferred to an independent board of directors, we can speak of the *public corporation* (cell (4,3)). A public corporation (or public company; used interchangeably) is thus similar to an entrepreneurial firm where the board stands for the entrepreneur(s), except that the hired capital is not private, but is owned by the state. Its similarity with the capitalist firm of the joint-stock type lies in dissociation of the functions of management and ownership of capital. However, in the joint-stock firm we have assumed that the entrepreneurial role is shared between stockholders (capitalists) and managers, while in a public corporation we assume that it belongs solely to the management. This requires an explanation. The differential treatment of the management's role in these two types of companies is explained by the greater influence that private stockholders exert on the policy of the joint-stock company than the state does on the policy of a public corporation. This is due to a greater awareness on the part of private individuals about their immediate economic interests. The shareholders' displeasure can quickly be transmitted to the management, either through the sale of shares in the stock market or through a change of the board at the general meeting.[5] The management of the joint-stock company can thus be viewed as partly autonomous in its decision-making and partly as a mere transmission of the capitalists on whose support it ultimately depends. In a public corporation the level of state interference in management is lower primarily because of the greater inertness of public bodies; state ownership is always, as Raymond Aron writes (1965, p. 161), an abstract ownership.[6] If, however, this changes and the board loses its autonomy, the firm by definition becomes state-socialist. There is, of course, always a grey area where it will be a matter of judgment to determine whether the entrepreneurial function is still performed by the management or has been taken over by the state. The same could be said for the relative importance of management and capitalists in joint-stock firms. It would clearly vary from firm to firm. Yet the ideal types that we have in mind should be clear.

An early example of the public corporation is provided by a practice that existed in Venice. Vessels built and owned by the state were leased at annual auctions. The state received an income derived from its ownership of assets (leasing of the ship), but it was not involved in management.[7]

The difference between a state-socialist firm and a public company is extremely important because, as we shall see later, these two types imply different modes of coordination of economic decisions (centralized vs. decentralized).[8] In order to differentiate them terminologically we shall use the term *public sector* when we speak of the segment of the economy composed of public companies, and *state sector* when we speak of the segment composed of state-socialist firms (or *state socialism*, if almost the entire economy consists of such firms).[9]

It is important to note that the state sector and the public sector do not coincide with firms in which the state holds, respectively, a majority or a minority stake. In distinguishing the state, public, and capitalist sectors we are concerned primarily with actual control over management of the enterprise (i.e., with entrepreneurship) and the type of coordination that exists among economic agents. Firms in which the state is simply one shareholder among many still remain, for all intents and purposes, capitalist joint-stock companies. If the state is a dominant shareholder, however, things are different. The state can then effectively exercise the entrepreneurial role, in which case there would be a gradual tendency for private shareholders—who no longer have a say in the management—to withdraw. The enterprise would become a state-socialist firm. Alternatively, the state may, despite its dominant shareholding, leave the managerial role to the board. We then have a public company (which can be majority or even entirely state-owned) whose coordination with the rest of the system follows a decentralized pattern.

As Table 1.2 shows, the situation with respect to entrepreneurship is fairly clear for a closed and for a joint-stock capitalist company. In the first, the entrepreneurial role belongs to capitalists; in the second, it is shared between capitalists and management. If the state is only one among many shareholders, the firm's position is not altered. But when state ownership becomes more important, there appear three claimants for the entrepreneurial role: the state, private capital owners, and the management. We have defined a pure public company as one in which management alone exercises the entrepreneurial role; in reality, there would be some sharing of this role between the shareholders (the state alone, or the state along with private capitalists) and the management, with the latter having the preponderant role. Finally, if the state entirely takes over the entrepreneurial function, we have a state-socialist firm.

Note that if a given amount of capital owned by the state were invested in fewer enterprises (which could then be completely controlled), the temptation (and the ability) of the state to implement centralized decision-making would be greater. Dispersion of state ownership thus appears to be one of the conditions for the preservation of a decentralized system.

It is incorrect to believe that the relation between the state and private capital is necessarily antagonistic. In fact, private capital often welcomes limited investment by the state: first, in order to dispose over more funds, and second, in the expectation that it might afford the firm more advantageous treatment (e.g., increased protection for its products, checks on domestic competition from other firms, more subsidies, etc.).

We have reviewed the four most important ways in which the process of production can be organized: capitalism, entrepreneurial system, state socialism, and public corporation. In all of them, workers are legally free, and exercise no entrepreneurial role. The difference distinguishing the systems resides in the position of capital. The ownership of capital conveys the ''right'' to organize the process of produc-

Table 1.2

Distribution of the Entrepreneurial Role

Type of firm	Capitalists	State	Management
With dominant private ownership			
Limited liability	+ +		
Joint stock	+		+
With dominant state ownership			
Public corporation	+	+	+ +
State socialist		+ +	

Note: + = Some entrepreneurial role.
       + + = Dominant entrepreneurial role.

tion in both capitalism and state socialism (although this right might be "diluted" in a joint-stock capitalist firm). The difference is that in one case capital is owned privately, and in the other, by the state. In the entrepreneurial and public corporation systems, capital, like labor, is hired, and the entrepreneurial role belongs to the management of the firm.

1.4 Workers as Entrepreneurs

Let us now consider forms of production in which workers exercise the entrepreneurial role. The most obvious is the one-man enterprise, where the same person works, uses his own capital, and decides about the organization of production. Examples are independent farmers, small-scale craftsmen, artisans, street vendors, doctors or lawyers working individually, consultants, and all self-employed members of liberal professions. These are *owner-worker firms*, with all three functions (labor, capital, and entrepreneurship) united in the same person. This type also includes all family-based enterprises. Another variant of the same type is a cooperative in which people pool their capital and labor and jointly organize production. We shall call this particular type *cooperative-1*. It represents an attempt to overcome problems inherent in small-scale production, while preserving the same organizational form. The most common examples are partnerships in the legal and medical professions. The self-employed and cooperative-1 types—which we combine under the title owner-worker firms—are shown in cell (1,2) in Table 1.1. The difference between cooperative-1 and the simplest capitalist firm lies in the absence of hired labor. The additional difference between cooperative–1 and a more evolved form of capitalist firm, like a joint-stock company, lies in the direct participation of owners of capital in the process of production.[10]

To the extent that cooperative–1 tends to use capital that does not belong to its members, it moves toward the *cooperative–2* type (cell (2,2) in Table 1.1). The pure cooperative–2 is where a worker or workers organize production, and carry it out by means of borrowed capital. An example would be individual taxi drivers who rent their cabs. The difference between cooperative–2 and the entrepreneurial firm lies in the absence of hired labor.

Finally, if capital is owned by the state, but is leased to workers who organize production, we have a *labor-managed firm* (cell (4,2)). Examples are labor-managed firms in Yugoslavia, the leasing of state-owned assets to workers in Hungary, and some cooperatives in socialist countries. The similarity with state socialism lies in state ownership of capital; the difference is the workers' entrepreneurship.

1.5 The Distribution of Income

The distribution of income that corresponds to each of the different modes of organization is shown in Table 1.3. The appropriation of income depends on modalities of ownership and the exercise of the entrepreneurial function. There is thus a direct linkage of ownership, entrepreneurship, and distribution of income.

1.6 Economic Coordination

The economic coordination of a system can be achieved by centralized means or it can result from the independent decisions of individual economic agents. Whenever the state organizes the process of production (i.e., the state is the entrepreneur), coordination must perforce be centralized. This is because the whole process is organized by a single entity—the state. In other words, there is only one economic agent in the proper sense, since the degree of decision-making autonomy of all other agents is nil. This one economic agent, the state, is generally associated with the center; hence we speak of centrally planned systems. It is important to notice, however, that the nature of the system is not altered if the state directs economic life (allocates inputs and outputs, sets production targets of the firms, etc.) at different levels, e.g.,

Table 1.3

Recipients of Income by Mode of Production and Type of Income

| | Type of income | | |
Mode of production	Labor	Capital	Entrepreneurial
Owner-worker	Workers	Workers	Workers
Capitalist limited-liability	Workers	Capitalists	Capitalists
Capitalist joint-stock	Workers	Capitalists	Capitalists/Managers
Cooperative–2	Workers	Capitalists	Workers
Entrepreneurial	Workers	Capitalists	Entrepreneurs
State-socialist	Workers	State	State
Labor-managed	Workers	State	Workers
Public corporation	Workers	State	Managers

central, provincial, or local.[11]

Whenever capital is privately owned, coordination is necessarily decentralized. This is so because ownership is dispersed. The owners of capital can either organize the production themselves or let different people bid for their capital. In either case, the number of independent economic agents (i.e., people who make production decisions) will be large and coordination will be decentralized.* Only if all the capital in a society were to be concentrated in a single owner would decision-making become centralized. All decisions would derive from a single agent. Consequently, as long as more than one agent owns the capital,

---

*And this does not depend upon the relative scarcity of capital with respect to labor. There is, in effect, an interesting relationship among (1) relative abundance of the factor of production, (2) exercise of the entrepreneurial function, and (3) private or state ownership of capital. If labor is more abundant than capital, the entrepreneurial function will tend to be taken over by the owners of capital. If capital is privately owned, coordination will be decentralized; if it is state-owned, coordination will be centralized. So far this is the "normal" situation. Now if we invert the relative scarcities of the factors of production, and let capital become the relatively more abundant factor, the entrepreneurial role will tend to accrue to labor. So long as capital is private, coordination will remain decentralized. But if capital becomes owned by a single agent (the state), coordination could become centralized—*even in the presence of a relative abundance of capital*—because the sole owner would be able to create artificial scarcity through monopolization. In the situation of scarcity he can control production—i.e., let capital be used only by those who agree to produce what they are told—and thus take over the entrepreneurial function and reimpose centralization. We therefore conclude that the root cause of centralization is to be found in a single agent's ownership of capital and not in the (natural) relative scarcity of a factor of production, or in the link between the (naturally) scarce factor and the entrepreneurial role.

the system based on private ownership of capital must be decentralized.

Table 1.1 shows, however, that the reverse does not follow: it is not true that decentralized coordination is possible only if capital is privately owned. Or, to put it differently, the existence of state ownership of capital is not a sufficient condition for centralization. If the state does not take over the entrepreneurial role, and instead hires out capital either to workers' cooperatives or to public corporations with autonomous management boards, coordination will be decentralized.

1.7 Firms' Maximands

In different modes of production, different social groups will exercise the entrepreneurial function. This implies that entrepreneurial agents in different arrangements will be motivated by the maximization of different objective functions. Capitalist firms will tend to maximize the return per unit of capital; entrepreneurial firms, the absolute amount of profits; labor-managed firms, income per worker; owner-workers, net income above the opportunity wage and return to capital; public corporations, managers' bonuses. These are, of course, stylized facts. The exact maximands will vary. The most important exception is that of a joint-stock firm, whose maximand, in reflection of the divided entrepreneurial role, will be some combination of return per unit of capital and managers' bonuses. The only type of firm for which no directly apparent maximand can be discerned is the state-socialist firm.

Differences in maximands imply differences in *equilibrium* prices in different decentralized modes of production.* Relative prices will differ even if the initial amounts of the primary factors of production are exactly the same. Allocations of the primary factors, production levels of different commodities, and the rates of investment and growth of the economies will differ. The conclusion that relative prices vary with the

---

*Theoretically speaking, equilibrium prices are indeterminate in a centrally planned system. Individual "departments" of the economy do not directly maximize an economic objective function. All of their parameters are determined by the central authorities. The firms may try to produce above the plan (if the managements try to please their superiors) or to produce below the plan (if they want to avoid high plan targets in the future). In either case a price structure does not spontaneously evolve as a result of individual decisions of economic agents. The price structure is determined by the central authorities in accordance with their objectives (e.g., perceived needs of the country, bargaining of different interest groups, etc.). Therefore, if we are interested in the relative price structure of a socialist economy, it is the objectives of the central authorities that are the relevant subject of study, not the reactions of the firms.

economic system is important first, because it directly translates underlying differences in the organization of production into the more tangible realm of prices, and second, it has obvious (and, so far, overlooked) implications for the theory of comparative advantage and international trade.[12]

1.8 Capitalist Economies

The actual organization of production in any contemporary society represents a combination of the different modes of production shown in Table 1.1.

> **Definition 4.** A combination of different modes of production will be called the *structure of production*.

In any given economy one mode of production will generally be the dominant mode. This mode will tend to impart its main characteristics to the whole system. In capitalist economies, the capitalist (joint-stock and limited-liability) mode is obviously the dominant mode. A growing share of state ownership in OECD countries since the end of World War II has made the public corporation the second most important mode of production. The percentage of gross domestic product (GDP) produced by the public and state sectors (excluding government services) in selected OECD countries ranges between 1 and 15 percent (see Table 1.4). France and Austria top the list, and the United States (predictably) is at the bottom. In most countries, the share of the public and state sector output in total GDP oscillates around 10 percent.

The links among public corporations themselves, and between them and the rest of the economy, are horizontal (decentralized). This is, first, because such coordination (as explained in section 1.6) is typical for public corporations anyway, and second, because the dominant, in this case capitalist, mode of organization tends to determine the prevalent type of coordination in the whole system.[13] Yet it cannot be unambiguously stated that, if the public sector were to become the dominant sector, the coordination would continue to be decentralized. In effect, as soon as the public sector starts to be regarded (by the government) as a single whole, and not as an agglomeration of individual enterprises, there is an implicit slide toward its transformation into the state sector. If the public sector is large and gets transformed into the state sector, the latter will inevitably imprint its characteristics on the rest of the

Table 1.4

Importance of Public Corporations and State Sector in Some OECD Countries*

	In terms of output	In terms of employment
High share (above 15%)†		
France (1982)	16.5	14.6
Moderate share (10 to 15%)		
Austria (1978–79)	14.5	13.0
Italy (1982)	14.0	15.0
France (1979)	13.0	10.3
New Zealand (1987)	12.0	
France (1973)	11.7	9.3
Turkey (1985)	11.2	20.0
United Kingdom (1978)	11.1	8.2
West Germany (1982)	10.7	7.8
United Kingdom (1983)	10.7	7.0
West Germany (1977)	10.3	7.9
United Kingdom (1972)	10.2	7.8
Sweden		10.5
Finland		10.0
Low share (5 to 10%)		
Portugal (1976)	9.7	
Australia (1978–79)	9.4	4.0
Denmark (1974)	6.3	5.0
Greece (1979)	6.1	
Norway		6.0
Canada		5.0
Negligible share (below 5%)		
Spain (1979)	4.1	
Netherlands (1971–73)	3.6	8.0
United States (1983)	1.3	1.8

Sources: See Annex 1.

*Excludes government services proper; i.e., includes only state-owned enterprises in commercial activities.
†According to share in output (whenever available).

economy, thus gradually turning it into a predominantly centralized one. In that sense, Hayek's (1944) contention that expansion of the public sector would lead to a centralized planning system is not implausible.

In less developed countries (LDCs) the importance of the public and state sectors is even greater (see Table 1.5). The reasons why the public sector has grown faster in these countries than in the developed capitalist part of the world (e.g., shortages of private capital and indigenous entrepreneurship, attempts to strengthen the unity of the country by centralizing capital resources at the state level, the desire to increase the power of political elites, etc.) are well documented. In a number of countries the share of output produced by the public sector exceeds 20 percent of GDP, and in some, where mining activities are particularly important (Zambia, Venezuela) or where pro-socialist policies have been followed (Guyana), it is even greater. It is low either in industrially very undeveloped countries (Nepal) or in traditionally anti-statist ones (Thailand, the Philippines). The economic coordination has generally remained decentralized, although in a number of the less developed countries the state sector has become the dominant sector in non-agricultural activities, thus bringing about centralized coordination. Examples include Cuba, Ethiopia, Vietnam, and Mongolia.

The example of the LDCs shows that the dominant mode of production in nonagricultural sectors determines the prevalent type of coordination among economic agents. In the case of developed countries this is less clear because the importance of agriculture in GDP is small anyway. In LDCs, however, even if agriculture contributes more than half of the national output, it is the type of organization in non-agricultural activities that gives its imprint to the whole economy. In other words, if most of the industry is made up of public-sector firms, while most of the agriculture is privately owned, it would be correct to concentrate one's attention mainly on the former, due to the essentially passive role of agriculture. There are two reasons for this. First, the political power is concentrated in cities, where the relevant political struggles also take place. Consequently, the economic activities carried out in cities and the welfare of the urban population are much more important, from the political point of view, than the activities and the welfare of the rural population. Second, the share of agriculture in the national economy shows a secular decline. Therefore, concentrating on nonagricultural sectors is tantamount to concentrating on sectors that will shape the national economy more and more decisively. The mode of organization existing there will be of far greater consequence than

Table 1.5

Importance of Public and State Sector in Some LDCs*

	In terms of output	In terms of employment
High share (above 15%)†		
Sudan (1980s)	40.0	
Zambia (1972)	37.8	
Guyana (1978–80)	37.2	
Venezuela (1978–80)	27.5	
Tunisia (1978–79)	25.4	
Singapore (1983)	25.0	
Guinea (1979)	25.0	
Chile (1981)	24.0	
Senegal (1974)	19.9	
Moderate share (between 10 and 15%)		
Taiwan (1978–80)	13.5	
Tanzania (1974–77)	12.3	
Bolivia (1974–77)	12.1	
Togo (1980)	11.8	
Ivory Coast (1979)	10.5	
India (1978)	10.3	6.0
Low share (between 5 and 10%)		
Niger (1980s)	10.0	
Sri Lanka (1974)	9.9	
India (1971–72)	9.4	
South Korea (1971–72)	9.1	
South Korea (1981–83)	9.0	7.0
Kenya (1970–73)	8.7	
Sierra Leone (1979)	7.6	1.3
Botswana (1978–79)	7.3	
Liberia (1977)	6.8	
Pakistan (1974–75)	6.0	2.8
Bangladesh (1974)	5.7	
Negligible share (below 5%)		
Thailand (1970–73)	3.6	
Paraguay (1978–80)	3.1	
Philippines (1974–77)	1.7	
Nepal (1974–75)	1.3	
Guatemala (1978–80)	1.1	

Sources: See Annex 1.

*Excludes government services proper; i.e., includes only state-owned enterprises in commercial activities.
†According to share in output.

the one existing in the countryside. The relative unimportance of the mode of production obtaining in agriculture is well illustrated by the state-socialist countries. In some of them (e.g., Poland) almost all of agriculture is privately owned. The dominant mode of production in the rural areas is not state-socialist, but either capitalist or cooperative-1. From our previous analysis, we know that these modes of production require decentralized coordination. Yet the dominant feature of the system is the centralized coordination dictated by the state-socialist sector's preponderance in nonagricultural sectors. Although the situation is not as clear-cut in some LDCs, due mostly to the size of the agriculture and thus to the percentage of the labor force that is not part of the system existing in the cities, the general principle is the same.

In addition to the capitalist, state-socialist, and public-corporation modes of production, there are also entrepreneurial and owner-worker firms in capitalist countries, both in developed and less developed ones. The importance of entrepreneurial enterprises is small. They are virtually nonexistent in less developed countries, while in developed capitalist economies they are limited to venture-capital firms.[14] Owner-worker enterprises are found commonly in agriculture and services. They are more important in less developed countries because they require very little capital, and there is no capital intensive production in agriculture to make small-scale enterprises uncompetitive. Further, in conditions of urban unemployment, owner-worker firms have more appeal, for whoever disposes of a minimum capital, than an unlikely employment in the "modern" sector. This explains their role as suppliers of different services, for example, street vendors, shoe-shine boys, food outlets, copy shops, etc. In effect, when development economists speak of the two sectors—modern and traditional—it is precisely the contrast between the capitalist and state-socialist types on the one hand, and the self-employment and cooperative-1 types (including family-based production) on the other, that they have in mind. It is the absence of hired labor, due to the paucity of capital, that differentiates the traditional sector from the modern. The importance of cooperative-1 firms in developed market economies is minimal, except in Italy (see Table 1.6). The proportion of the self-employed, on the contrary, seems to be on the increase. In the United States about 12 percent of the male labor force, and about 7.5 percent of the female, were classified as self-employed in 1982; these figures were 10 percent and 5 percent respectively in the late sixties.[15] In the United Kingdom the self-employed accounted for 10 percent of total employment in

Table 1.6

Importance of Cooperative-1 Type in Selected Capitalist Countries

	Number of firms	Employment
Italy (1982)	11,203	527,900
United Kingdom (1984)	911	8,800
France (1982)	1,080	35,000
Spain (1984)	1,179	36,500
Belgium (1984)	61	5,634
West Germany (1984)	19	700
Netherlands (1984)	60	2,000
Sweden (1984)	114	4,000

Sources: See Annex 1.

1985, and only 8 percent in 1971.[16]

Labor-managed firms are virtually nonexistent in either developed or developing capitalist economies. This is due to the fact that, whenever the state has ownership over the means of production, it either takes over the role of the entrepreneur or leaves this function to an autonomous management board. In other words, labor assumes the entrepreneurial role only when it also possesses capital (as in owner-worker firms), which implies that *the entrepreneurial role is generally associated with the ownership of capital.*[17]

1.9 Socialist Economies

In the socialist countries, it is the state-socialist mode of production that is dominant. The importance of this mode by far exceeds that of any other. As Table 1.7 shows, the share of the state sector in socialist countries varies between 65 percent and almost 100 percent of GDP. It is lowest in Hungary and highest in Czechoslovakia and East Germany. It is notable that in none of these countries (except in China for which the data refer to industrial sector only) does the state sector produce less than two-thirds of total output. To some extent state-socialist countries offer a mirror image of the capitalist (OECD) economies. While in the latter the state sector is the marginal producer with about 10 percent of GDP, in socialist countries it is the entire nonstate sector (including capitalist, owner-worker, and labor-managed firms) that is the marginal producer, contributing about 20 percent of GDP.

Table 1.7

Importance of the State Sector in Some Socialist Countries*

	In terms of output	In terms of employment
Very high share (above 85%)†		
Czechoslovakia (1986)	97.0	
East Germany (1982)	96.5	94.2
Soviet Union (1985)	96.0	
High share (70–85%)		
Poland (1980)	83.4	73.4
Poland (1970)	82.2	68.0
Poland (1985)	81.7	71.5
China (1980)	78.7	
China (1982)	77.8	
China (1984)	73.6	
Hungary (1975)	73.3	70.9
Moderate share (50–70%)		
Hungary (1980)	69.8	71.1
Hungary (1984)	65.2	69.9

Sources: See Annex 1.

*Includes the government proper.
†According to share in output.

In the table we have included kolkhozes and their equivalents in East European countries as part of the state sector. This requires an explanation, both because of the Soviet official classification of kolkhozes as cooperatives and because the kolkhoz type offers an interesting combination of features specific to different modes of production. The distinguishing features of kolkhozes are: production quotas (deliveries), with prices for the deliveries determined by the state; at times the absence of the right to leave the kolkhoz; and peasants as claimants of the residual income, with each paid according to his number of workdays. Absence of free mobility of labor (which was the case in the Soviet Union until the mid-1950s) makes kolkhozes similar to the state-regulated systems with bonded labor. When labor is legally free, the fact that kolkhozes cannot choose their own product mix, are subject to specific production targets (in terms of quantity), and are an integral part of the comprehensive central planning system clearly puts them into the category of state-socialist firms. Finally, as labor is the residual income claimant, there are also some elements of the labor-managed system. However, since our main classification principle is the effective exercise of the entrepreneurial role, and in the kolkhozes the entrepreneurs clearly are not the farmers, but the state, kolkhozes must be included among state-socialist firms. On the other hand, the work performed by farmers (members of a kolkhoz) on their own private plots represents self-employment.

Coordination in the state sector is carried out by centralized means. The autonomy of individual firms is negligible or nil. In effect, there is only one economic agent in the proper sense: the state.

Owner-worker firms exist in state-socialist countries mostly in agriculture and services (e.g., farming in China and Poland). This is the only important form in which privately owned capital (or *de facto* privately owned as in China) appears. The labor-managed mode of production similarly exists in agriculture, and increasingly in industry. Such evolution is particularly apparent in Hungary, where the system of leasing state-owned capital to groups of workers was started.[18] It is noteworthy that there has been an incipient pressure in socialist countries (in some segments of the economy) toward a shift from the state-socialist to the labor-managed system. This pressure has appeared in the open every time popular movements disaffected by the centralized system have gained in importance. The reasons for this attraction seem clear. If state ownership over the means of production is taken as given, the attractiveness to workers of a situation where they are also entrepreneurs is obvious. An additional reason derives from the decentralized structure of coordination implied by the labor-managed system. These points will be taken up more fully in chapter 5.

The public sector is only emerging. It is somewhat difficult to estimate its numerical importance since it is composed of (nominally) state-socialist firms that have been successful in weaning themselves away from state influence (or were allowed to do so). As the central control over state firms is relaxed (e.g., by having plan targets cover only a portion of production), enterprises slowly evolve into public corporations. In several socialist countries (Poland, Hungary, China) this sector is probably the most rapidly growing one, although it is still small in absolute size and legally indistinguishable from the state sector. But, as it expands, it is not unreasonable to expect that even its legal status will come to be different from

---

Table 1.8

Relative Importance of Different Modes of Production

Ranking	Capitalist countries	Socialist countries
1	Capitalist	State-socialist
2	Public corporation	Owner-worker†
3	State-socialist*	Labor-managed
4	Owner-worker†	Public corporation
5	Entrepreneurial	Capitalist

*The relative importance of the state-socialist mode is fairly high if we also include "the government activities proper"—not only the law-and-order type, but also health, education, and other social services. It is much less if we only include commercial undertakings.

†Includes self-employment and Cooperative–1.

---

that of the classic state-socialist firm.

There are also capitalist enterprises in state-socialist countries. Their growth is circumscribed by statutory limits on the number of workers they can employ. The importance of this type is accordingly fairly small.

1.10 The Structure of Production and the Legal Framework

The objective of the above analysis has been to place different modes of production into a unified framework and to show that different modes coexist in every economy. This suggests that different countries with different modes of production varying in importance can be regarded as lying along a continuum rather than as representing discrete cases.

Table 1.8 shows the stylized ranking of different modes of production according to their share in total employment or output in capitalist and socialist economies. In other words, it gives a typical structure of production.

The main difference among the countries, and the one where we can indeed speak of a discrete change, resides in the mechanism used for the coordination of economic decisions. The dominant mode of production (or, more exactly, the coordination mechanism proper to that mode of production) determines the ruling system of coordination among the agents. This is why we find that in capitalist countries decentralized coordination extends to the enterprises with a large share of state capital; and that in socialist systems decentralized coordination

Table 1.9

Alternative Modes of Organization of Production in Poland and the United Kingdom (in % of total employed)

			Labor		
			Entrepreneurial role	Hired out	Total
		Poland (1985)			
Capital ownership	Private:	Entrepreneurial role	23.1	5.4	28.5
		Hired out	0	0	0
	State:	Entrepreneurial role	—	71.5	71.5 ⎫
		Hired out	0	0	0 ⎬ 71.5
	Total		23.1	76.9	100
		United Kingdom (1985)*			
Capital ownership	Private:	Entrepreneurial role	10.2	62.5	72.7
		Hired out	0	0	0
	State:	Entrepreneurial role	—	22.0	22.0 ⎫
		Hired out	0	5.3	5.3 ⎬ 27.3
	Total		10.2	89.8	100

Note: Original data and sources in Annex 1.

*The data given here differ from those in Table 1.4. The latter include only employment in the state sector outside the government proper, i.e., in the commercial activities provided by state-owned enterprises. This would be equivalent to the ratio between employment in public corporations (5.3%) to total nongovernment employment (78%), i.e., 6.8%, which is close to the value for the UK given in Table 1.4.

can appear only sporadically, and only to the extent that centralized decisions leave room for it. The dominant mode of production defines the economic system.

Definition 5. By an *economic system*, let us say the capitalist system, we mean a system in which the capitalist mode of production is the dominant one. The importance of other modes of production can vary to some extent without affecting the system.

Table 1.9 illustrates the importance of alternative modes of production in Poland and the United Kingdom in terms of total employment. From the structure of employment in Poland and the United Kingdom we can readily make comparisons between the two countries. The dominant modes of production are obviously state-socialist (71.5 percent of total employment) in Poland and capitalist (62.5 percent) in Great Britain. But the data allow us to make further comparisons: association of labor with entrepreneurship is more than twice as frequent in Poland as in the United Kingdom: 23.1 percent versus 10.2 percent. This is mostly due to the high share of the agricultural labor force in Poland (self-employed). The absence of (*de jure*) public corporations in Poland contrasts with more than 5 percent employed in them in the United Kingdom. The most revealing, however, is the comparison between percentages of people who work on privately owned and state-owned capital. In Poland the ratio is 2.5 to 1 in favor of state-owned assets (71.5 vs. 28.5 percent); in the United Kingdom it is exactly the reverse: the ratio is 2.7 to 1 in favor of people working on privately owned assets (72.7 vs. 27.3 percent). The coordination in Poland is predominantly centralized: more than 70 percent of the employed work within the centrally planned system; in the United Kingdom this proportion is only 22 percent. The relation illustrates the fundamental difference between the two systems.

Using the same approach we can study the evolution of the structure within a country. We shall consider the example of Great Britain, a country that has undergone significant structural changes in the last fifteen years. Table 1.10 gives the structure of employment in 1971 and 1981.

Between 1971 and 1981 there were 2.6 and 0.4 percentage point declines in the share of those employed in the capitalist and public corporation sectors respectively. These were matched by almost equivalent (2.5 and 0.5) increases in the state sector (mostly central government and local authorities services) and self-employment respectively. The percentage of people who work on state-owned capital expanded from 27.4 to 29.5 percent (even though the share of the public corporations went down). The effects of the "Thatcher revolution" are apparent when we consider the structure of employment in 1985 (in Table 1.9). Between 1981 and 1985, the only sector to register a decline in the share of employment was that of public corporations[19]: it decreased by 2.5 percentage points. Employees released by that sector joined the ranks of the self-employed (1.6 percentage points increase), or were absorbed by the capitalist (+0.6 point) and state (+0.3 point) sectors. Consequently, the two most significant effects were a substantial increase in the numbers of the self-employed and a decrease in public corporation employment. It is significant that employment in the capitalist sector increased only slightly in the 1981–85 period (and was still

Table 1.10

The Structure of Production in the United Kingdom, 1971 and 1981 (in % of total employed)

			Labor		
			Entrepreneurial role	Hired out	*Total*
	1971				
Capital ownership	Private:	Entrepreneurial role	8.1	64.5	72.6
		Hired out	0	0	*0*
	State:	Entrepreneurial role	—	19.2	19.2 } 27.4
		Hired out	0	8.2	8.2
	Total		8.2	91.9	100
	1981				
Capital ownership	Private:	Entrepreneurial role	8.6	61.9	70.5
		Hired out	0	0	*0*
	State:	Entrepreneurial role	—	21.7	21.7 } 29.5
		Hired out	0	7.8	7.8
	Total		8.6	91.4	100

Sources: See Annex 1.

lower than in 1971 by 2 percentage points), and that employment in the state sector continued to expand. The ratio between the number of people working on privately owned vs. state-owned capital increased from 2.4:1 in 1981 to 2.7:1 in 1985.[20]

Definition 6. The set of legal rules that defines different modes of production is called the *legal framework*. The legal framework defines property rights. The legal framework is always the result of a social contract. It is irrelevant for our purpose whether the contract is enforced by force or reflects a general social consensus about the forms of ownership.

The legal framework consists of a set of laws and rules that govern the property relations in the process of production. It is a legal *pendant* to each economic mode of production. By treating some types of property relations more favorably, and/or by banning or discouraging others, the legal framework strongly affects the structure of production. To explain: If (as was the case in Britain until the mid-nineteenth century) a special Act of Parliament is needed to form a joint-stock company, it can be reasonably argued that the legal framework is not particularly congenial to this type of arrangement. Similarly, if the law forbids private individuals from hiring more than a prescribed number of workers, or if the constitution prohibits the government from undertaking commercial activities, these regulations will stifle the growth of capitalist or state-owned enterprises respectively. Finally, if slavery is outlawed, no mode of production with unfree labor will legally exist. It is then evident that there must be a relationship between the legal framework and the actual structure of production, although this relationship is not uniquely determined. That is, the importance of different modes of production may, within some limits, vary inside a given legal framework. The change in the structure of employment observed in the example of Great Britain may thus be formally decomposed into a technology- and taste-determined part (exogenous from our point of view) and a change induced by the alteration of the legal framework. It is this second element—namely, the privatization of public corporations—that constitutes the ''Thatcher revolution.'' Similarly, it is obvious that if statutory limits on capitalist firms were lifted in Poland, that sector would expand.[21]

It is the role of economists to study how changes in legal regulations affect the importance of different modes of production, and how the new structure of production alters the demand for labor and capital and leads to changes in the output mix of the economy. In other words, if a change in laws stimulates the expansion of a mode of production that is generally used in small-scale production, this will have definite effects both on the demand for factors of production (e.g., demand for labor may go up) and on the composition and the size of final output. We thus clearly grasp the linkage among (1) the legal framework, (2) the structure of production (i.e., relative importance of different modes), and (3) the demand for factors of production and the composition of final output. The last element, of course, determines income distribution, which—through effects on consumption and saving patterns—in turn influences production decisions.

Chapter 2

The Economic Role
of the State

"What is interventionism? Interventionism means that the government does not restrict its activity to the preservation of order, or—as people used to say a hundred years ago—to the 'production of security.' Interventionism means that the government wants to do more. It wants to interfere with market phenomena."

—Ludwig von Mises*

2.1 Definition of State Interference

Definition 7. *State interference* in the economic sphere is defined as all state actions that (a) limit the area of choice of an economic agent, or (b) make an economic agent not entirely responsible for his decisions, or (c) impose any cost on the agent's activities. This definition of state interference means that the economic role of a "non-interfering" state is limited to the enforcement of the legal framework, that is, to the protection of a given set of property rights. Any other action falls under the category of interference.

Note that we use "interference" only as a technical term that is value-neutral; it is not intended to convey disapprobation or approbation of a specific state activity.

Category (a) includes all actions that fix or limit supply, demand, or

---

*Ludwig von Mises, *Economic Policy*, Regnery/Gateway, 1979, p. 39.

prices of products and factors of production. These actions (e.g., price controls, exchange-rate regulations, ceilings on interest rates, etc.) may be pervasive, affecting a number of economic agents, or specific (e.g., fixing the price of a particular product). The distinguishing characteristic of these actions is that the range of choice for an economic agent becomes more restricted.* It is important to note here that the choice becomes more restricted *within* a given legal framework. If, according to the legal framework, there is a limit on the size of a particular type of organization (e.g., a ceiling on the number of workers that a capitalist firm can employ under the system of state socialism), this limit is not to be treated as state interference, but as a part of a given bundle of property rights.† Only under such a definition of a legal framework can we allow for (different) specific legal frameworks and avoid introducing an *atemporal* given legal framework. To explain: Consider a legal ban on slavery. If this were no longer to be treated as a part of a given legal framework, specific to some societies, but as state intervention, we would lose any operational concept of real state interference (namely, actions affecting prices, supply, demand, etc.). Differently, if we assumed that the legal framework *must* include freedom of labor and private property of capital, we could indeed treat any infringement on private disposal of assets as state interference, but we would be committing the gross error of erecting one historical type of property arrangement into a "natural" type; for the greater part of human history is characterized by the existence of legally unfree labor[1] and/or state (or communal) property of capital. These arrangements are consequently no less "natural" than the capitalist arrangement. We must

---

*More restrictions need not necessarily imply a less favorable position for a firm. On the contrary, state fixing of prices of some inputs, or foreign exchange below market level, can be very profitable for a firm. However, since trade must legally be conducted at these prices, the range of choice (even for the firms that gain from the regulation) is more restricted.

†Constraints placed on the development of some types of economic organizations by the legal framework are not limited to state socialism. Up to 1852 cooperatives were not legal in England. This was based on a law that banned all organization in restraint of trade, and cooperatives were considered to be this, since they were often founded with the objective of circumventing the market mechanism and replacing it with direct distribution. Similarly, until 1852 a limited-liability company could be created only by an express Act of Parliament (i.e., Parliament needed to give a dispensation that shareholders would not be responsible for the firm's debts with all of their private assets). In the Middle Ages guilds tried to prevent transformation of artisans into capitalists by limiting the number of apprentices.

thus take the legal framework—i.e., the acceptable types of property relations in a society—as given, and examine the changing role of the state within such a legal framework.

Category (b) includes all state actions that are related to direct subsidization (bail-outs of firms) and protection from internal or external competition. Subsidization of firms differs from interference (a) in the following sense: even if an agent is entirely free to make all of his own economic decisions, he may not have to face the consequences of these decisions. In practice, however, the two categories are often related. Price limits on particular goods, imposed for "social reasons," are often accompanied by subsidies. This category of state actions is responsible for creating the phenomenon of "soft budget constraint."[2] Protection similarly represents a legally enforced subsidization.

Category (c) includes all direct and indirect taxation of economic agents. These are costs imposed by the *fiat* of the state, belonging, strictly speaking, to the realm outside the economic sphere. Their distinguishing feature is that the payment is not the result of a free commercial contract between the enterprise and the state, but is a legally enforced obligation.

It is important to note that state interference, as defined here, is methodologically different from the issue of discretionary or uniform treatment of economic agents. State interference exists, in effect, regardless of whether all economic agents are treated equally or not. Differential treatment of some agents may introduce specific constraints (or advantages) for them, but does not, in itself, constitute the essence of state interference in the economic sphere.[3]

2.2 Mode of Production and State Intervention

A given legal framework and structure of production may be compatible with different degrees of government interference. For example, a predominantly capitalist economy can accommodate very little state interference (e.g., limited only to tax collection) or very extensive interference (e.g., price and wage controls, fixing of the exchange rate, etc.). The decentralized decision-making nature of the system is not affected. Even in the face of very strict state regulations, the system in which private property of capital is dominant remains a system that relies on decentralized coordination. (This goes back to our finding in 1.6 that all systems with private ownership of capital must be decentralized.) The set of permissible actions for each agent can be severely restricted, but his autonomy remains. There is no plan that purports to dictate his *behavior*. An extreme situation could arise if, within the framework of a capitalist system, the state resorted to such detailed regulations that most of the variables relevant for economic agents were fixed by the state. It might be thought that in such an extreme case the firm would be in a position no different from that of a firm in state socialism; but this is not true. The still existing autonomy of the capitalist firm implies that its right to go out of current business and/or to start a new one will be retained. Only if the autonomy is lost, and the firm loses its legal personality, will the decentralized system of coordination be replaced by the centralized.

Visually, we can conceive the absence of any state interference as being a situation in which an economic agent's area of permissible decisions is constrained only by the existing legal framework (e.g., if slavery is not legally recognized, the firm cannot make workers slaves even if they agree). As the role of the state expands, the area shrinks, and the set of permissible decisions is no longer delineated only by the legal framework, but also by the economic decisions of the state. Finally, under central planning, the area collapses into a single point—predetermined by the state—and choice for individual agents disappears.[4]

It is important to note that the discussion about the relationship between individual economic agents and the state applies equally to private and to state-owned firms. For example, the previous two paragraphs, where we sketched how a possible transition from decentralized to centralized coordination could take place, apply equally well to public corporations (with complete state ownership) and to private firms. Whether the state appears only as one among many shareholders in a private firm, or is the majority shareholder (as in a public corporation), or finally rents out capital to workers (as in the labor-managed system), the coordination among the firms remains decentralized.[5] Consequently, *economic interference of the state must be defined not in terms of the ownership of assets, but in terms of state actions that affect economic agents, regardless of the form of ownership of their capital.*

The link that exists in most people's minds between state interference and state ownership is based on the observation that the state intervenes (in the sense of points (a)–(c)) in the economic affairs of the firms in which it holds a majority stake or full ownership more frequently than in the affairs of privately owned firms. This is to be expected, since in many instances firms were nationalized precisely in order to serve some noneconomic objective, and accordingly interference of types (a) and (b)—the two being linked—is more common.

However, when the question of reduction of the role of the state in the economy is raised, it is methodologically necessary to keep the issue of interference separate from the issue of ownership. In other words, when we speak of interference, it is types of actions like (a)–(c) that we must address, not particular forms of ownership. Forms of ownership (modes of production) define the structure of production; they do not determine the level of state intervention.[6] Yet it is also correct—if the incidence of state interference is greater in state-owned enterprises—to address the question of interference from a different perspective, and to argue that if the state sector were eliminated, the likelihood of state interference would be lessened. One must, however, be careful to distinguish what really constitutes state intervention from a tactical consideration, namely, where it is more likely to be exercised.

We can illustrate the importance of distinguishing between state interference and the mode of production using the example of kolkhozes, discussed in section 1.9. Since the entrepreneurial role in kolkhozes is exercised by the state, we have included them in the state sector. However, if only one of the key characteristics of kolkhozes were changed, namely if state determination of the product mix and compulsory selling quotas were eliminated, kolkhozes could be reclassified as labor-managed firms (paying a rent to the state for the use of land). Even if the prices of their output were determined by the state, this would now amount to type (a) state interference exercised under conditions of *decentralized coordination*. This, however, is not a mere question of reclassification, for, in effect, a fundamental change would have taken place. Farmers would have become free to determine their own product mix; their production decisions would be based on their own objective function, maximized under conditions of externally fixed prices; and they would engage in developing horizontal links—a key ingredient of decentralized coordination—with other firms in the economy. They would thus effectively take over the entrepreneurial role. Even if some state intervention remained, the mode of production would have changed.

2.3 State Intervention in Capitalism and Socialism

We have shown how a given mode of production (e.g., capitalist) can be compatible with different levels of state intervention. However, this is not true for all seven modes of production (with legally free labor) displayed in Table 1.1. More specifically, this is not true for the state-socialist mode of production. To illustrate this point, consider the difference between the two pure cases, the capitalist and the state-socialist system, where *all* the firms in an economy are assumed to be respectively capitalist and state-socialist. Our matrix in Table 1.1 thus collapses to only one cell: cell (1,2) for the capitalist, and cell (3,2) for the socialist economy.

The state can exercise vastly different degrees of interference in the capitalist economy without the economy ceasing to be a capitalist one. The state can determine some prices, or leave all of them free; it can bail out some private firms, or let them go bankrupt; it can impose different tax rates on different products or firms, or impose no taxes at all. Private entrepreneurs/capital owners may in one case have to operate under very stringent regulations (their range of free decisions being accordingly curtailed), while in another case there may be no restrictions at all. However, in all these cases, the economy remains capitalist. The essence of the capitalist economy, as we defined it (entrepreneurship exercised by private owners of capital who hire labor, and decentralized decision-making), is not altered. Indeed, this is something strongly confirmed by experience: This is why we can speak of a regulated economy like England's after World War II, and Hong Kong's, as belonging to the same species. It also shows why those who argued that Keynesian macroeconomic policies (as distinct from state ownership of capital and expansion of the public sector; see section 1.8, above) would ultimately lead to a change in the essential mode of production (destruction of capitalism), were wrong.

The situation differs with state socialism. In state socialism in its pure form *all* economic decisions are taken by the state. As we pointed out above, there is *de facto* only one economic agent: the state. The range of freedom for individual enterprises is nil: all the decisions about prices, quantities of inputs and outputs, firms' suppliers and buyers have already been taken. Individual enterprises are simply sub-departments of the national economy (part of a single workshop, *mas-terskaia*, as Bukharin and Preobrazhensky called it). They are production units that transform specified quantities of inputs into specified quantities of outputs. Pure state socialism is therefore compatible with only one level of state intervention in the economy: the extreme level. When some controls over inputs or outputs of individual firms are relaxed, the firms—as individual entities—spring back to life. They are no longer subdivisions of a single whole. As soon as this happens the mode of coordination among the firms tends to become decentralized. The importance of this development cannot be exaggerated. The firms no longer call the center if the delivery of inputs is late; they try to find other firms that can supply the goods, or promise higher prices to suppliers. Horizontal links among the firms[7] immediately replace vertical (center–enterprise) coordination.

Thus, *any relaxation of state control in state socialism implies an inexorable movement toward the system of decentralized coordination. The nature of the system changes in an essential way.*

The implication of this finding is twofold. (In this we foreshadow fuller discussion of this topic in chapter 5.) First, changing the level of state intervention in state socialism is more difficult than it is in capitalism, because such a change directly affects the dominant mode of production. This is not an abstract statement, since each given mode of production implies a particular distribution of economic and political power. This is particularly true for the dominant mode of production because it determines the overall structure of economic and political power in the system, with the result that the people who have the greatest stake in its preservation are also the most powerful. Their resistance to a change of the dominant mode of production (which entails the replacement of one economic system by another) will be far greater than their resistance to some redistribution of power within the *same* system. In general, the power elite may be persuaded or coerced into giving up some power or losing some economic benefits, if this is required to get the system "unstuck." They will do so more easily if they have grounds to believe that they will be able to retain the upper hand after the reforms, precisely because the essence of the system will not be altered, even if their power is reduced. Loss of some power thus appears to be a condition for the survival of the elite. But it would clearly take more coercion to get those who profit most from the system to relinquish all power, and this is precisely what a shift to a new economic system implies.

The second implication relates to the likely directions of change. In the capitalist system, as we saw, change takes place by varying the role of the state within the same dominant mode of production. This is impossible in state socialism. Change there may consist either in (1) replacing the state socialist mode with a new dominant mode of production, compatible with the preservation of state ownership over the instruments of production, viz., in a movement toward a labor-managed or public-corporation type of system,[8] or in (2) allowing greater scope for other nondominant market-oriented modes of production. The second option is more likely. This is so because it need not entail any relaxation of state control in the greatest part of the economy. In other words, it would "rock the boat" least. The coordination would still remain centralized, and the dominant mode of production would remain state socialist. But, in order to introduce some flexibility, the "individual" private sector (i.e., owner-worker firms) as well as labor-managed and capitalist forms may be permitted to a greater extent. They will not be under the writ of the central authorities. Their role, however, must forcibly be limited so that they do not threaten the dominant role of the state sector. This would be an embryonic form of a state-socialist mixed economy. Consequently, the movement from the state-socialist system toward a full-fledged public-corporation or labor-managed system appears unlikely, since it entails the substitution of one dominant mode by another, and, accordingly, a change from centralized to decentralized coordination along with a significant redistribution of political and economic power.

To sum up: In capitalist systems liberalization generally takes place by reducing the role of the state within an unchanged legal framework. In state-socialist systems liberalization takes place by allowing greater scope for other nondominant forms of production, that is, by partly altering the legal framework.[9]

2.4 What Is Liberalization?

We define liberalization of the economy as any move that reduces the role of the state in accordance with rules (a)–(c). By implication, such a move must include at least one of the following: (a) increased autonomy of economic agents, (b) their greater responsibility, (c) reduction in taxation.

In discussions about liberalization, the following dichotomies are often used: market/plan, decentralization/centralization, public/private, and competition/monopoly. Our previous analysis enables us to define these terms more strictly.

A *market economy* is an economy in which coordination among economic agents is achieved principally by decentralized means. A *planned economy* is one in which coordination is achieved by centralized means. There is an absolute correspondence between the market/plan and decentralization/centralization dichotomies, and, accordingly, we can replace "decentralized" and "centralized" in

Table 1.1 with "market" and "plan."

The *private* and *public* sectors are defined according to the ownership of assets. When capital is owned by the state, we can have either a public or a state sector with, respectively, decentralized and centralized coordination. There is, accordingly, no one-to-one relationship between ownership and the form of coordination.

Because liberalization increases the autonomy of economic agents, liberalization is synonymous with a greater role for the market mechanism (decentralized decision-making). It is, therefore, correct to treat liberalization and an increased market orientation as the same thing. On the other hand, it is theoretically incorrect to identify privatization (which is a change in the legal framework) with liberalization of the economy, because privatization can coincide with an increased involvement of the state in the economy in the sense of points (a)–(c). However, if there is empirical evidence that state interference tends to be greater in the public sector, privatization may in effect be expected to result in a liberalization of the economy. It is important to keep in mind that liberalization and privatization are not identical, while increased market orientation and liberalization are.

A final point concerns the relationship between *perfect competition* and *monopoly*. To some extent (but *only* to some extent) they coincide with the previous market/plan dichotomy. If perfect competition were the only possible outcome of a liberal economic order, and monopoly the only outcome of a planned system, the correspondence would be total. But, while the second part of the statement is true, perfect competition is only one of many possible results of a liberal economic order. In other words, free competition, with an absence of government intervention in economic life (i.e., *laissez-faire*), is almost as likely to generate a perfectly competitive solution as it is to generate monopoly or oligopoly.*

This contradiction between a liberal economic order and a perfectly competitive market was forcefully stressed by Karl Polanyi (1957, p. 148):

> Whether workers' associations to raise wages, or trade associations to raise prices were in question, the principles of *laissez-faire* could

---

*A note on definitions. *Free* competition implies an absence of legal hindrances to entry and exit, free production and pricing decisions of the firms. *Perfect* competition requires that demand and supply functions faced by firms be perfectly elastic: each firm is a price-taker in all of its markets.

be obviously employed by interested parties to narrow the market for labor or other commodities. It is highly significant that in either case consistent liberals from Lloyd George and Theodore Roosevelt to Thurman Arnold and Walter Lippmann subordinated *laissez-faire* to the demand for a free competitive market; they pressed for regulations and restrictions, for penal laws and compulsion, arguing as any "collectivist" would that the freedom of contract was being "abused" by trade unions, or corporations. . . . Theoretically, *laissez-faire* or freedom of contract implied the freedom of workers to withhold their labor either individually or jointly, if they so decided; it implied also the freedom of businessmen to concert on selling prices irrespective of wishes of consumers. But in practice such freedom conflicted with the institution of a self-regulating market, and *in such a conflict the self-regulating market was invariably accorded precedence.*

The contradiction is particularly pertinent if we stay within the confines of a single national economy. If we broaden our gaze and look at the world as a whole, the likelihood of the emergence of a monopoly— which would need to be global or at least regional (if there are high transportation costs)—in a fully liberal system, characterized by free circulation of goods, labor, and capital, is significantly less. The problems of combination, on such a vast scale, with a much greater number of agents, are incomparably more difficult. Yet, theoretically, the possibility that even *laissez-faire* on an international scale results in a monopolistic or oligopolistic solution, cannot be ruled out.

A planned system, on the other hand, is consistent only with monopoly. Even if there are several enterprises producing a given good, they will not compete with each other, since their output and prices are determined by a higher authority.[10] Further, since a planned economy is a system defined at the national level, where foreign trade appears only as a residual (exports as a vent for surpluses, imports only of goods that are not produced at home), trade cannot be allowed to infuse some competition into the system and upset the realization of planned targets. As a standard practice, domestic and foreign prices, in areas where there may be some potential competition, will be equalized. The costs (or profits) of such an operation will be reflected in a special state price equalization account: prices in the domestic economy will thus be completely insulated from international prices.

Chapter 3

The Efficiency of Different Modes of Production

"It thus comes about that, in competition one with another, men look both to their own advantage and to that of the public; so that in both respects wonderful progress is made."—Machiavelli*

The objective of this chapter is to compare different modes of production, in order to identify essential types of obstacles to efficiency in each of them.

We define efficiency as the ability to produce maximum utility out of a given quantity of inputs (disutility). As the level of utility is proportionally related to the level of output produced, we can redefine efficiency to mean production of maximum goods and services with a given quantity of inputs (including disutility of labor).[1] The determination of the conditions for the efficient use of resources proceeds according to the following steps.

> **Statement 1.** The condition for maximum efficiency in the use of a nonanimate resource (capital or land) is that the legal owner of the resource be the one who decides on the use of the resource and appropriates all the returns from its use.

Statement 1 derives from the assumption of rational behavior. For, even if there are alternatives that, if selected, would have yielded a

---

*Niccolo Machiavelli, *The Discourses*, ed. Bernard Crick, Pelican Books, London, 1976, p. 280.

superior return to the owner, we must assume that (i) if the owner knew about all the alternatives, he decided not to select the one that eventually proved to be the most profitable, due to his subjective estimate of risk (his preferences) at the time of decision-making, or (ii) he ignored the alternatives. Ignorance does not affect the validity of Statement 1 since it is, like differential estimates of uncertain returns between different people, already included in the original conditions of the decision-making. What we are saying in effect is that, *given* his knowledge, his preferences, and his estimate of risks involved at the time of decision-making, the economic agent will always select the alternative that maximizes his utility (again, as estimated at the moment of decision-making).

Corollary of Statement 1. Resources are used inefficiently whenever (a) the agent who owns the resource does not decide on its use, and (b) the owner of the resource does not alone appropriate the return. Externalities and taxation are examples of (b). Externalities, however, are the product of badly defined (or undefined) property rights. They disappear if property rights are sufficiently comprehensive. But the existence of taxation, and thus of the state alone, implies inefficiency in the use of the resources.[2] This also means that one of the ways that the state intervenes in economic life—our point (c) in Definition 7—necessarily generates inefficiency. It may be noted that in the case of both externalities and taxation, a wedge is introduced between the actual yield of the resource and the return appropriated by its owner. As is well known from the property rights literature, the existence of a wedge must lead to inefficiency in the use of resources.

Diffuseness of ownership, and the absence of an unambiguous *concrete* title-holder to a resource, is the main reason for inefficiency in the use of state-owned capital. If the state were a well-defined homogenous group of people whose income derived directly from the use to which the capital is put, there would be no difference between direct capitalist ownership and state ownership. However, the income of people who (in the name of the state) decide about the use of capital (as in state socialism) or leasing of capital (as in the case of public corporations and labor-managed firms) is not directly related to the return generated by that particular investment. In other words, the link between the decision and the reward is severed. The reasons for inefficiency will be greater in state socialism than in the two other systems with state ownership of capital, because in the former the entrepreneurial function is also taken over by the state. The same type of weak relationship between decision-making power and the appropriation of returns extends here to the entrepreneurial income. Absence of responsibility (positive or negative) for economic decisions, due to the diffuseness of "owners" of capital, is the primary reason for the economic inefficiency of systems based on state ownership of capital.[3]

The state-socialist type of organization will also be inefficient because of the information *lacunae* common to all centralized systems. As shown by Hayek (1945, pp. 519–30), economic knowledge is always of a dispersed nature; it is the knowledge of specific opportunities of time and circumstance. Such "grass-roots" knowledge is available only to direct participants and not to central bodies. Accordingly, a number of opportunities will go unused in any centralized system, because of the inadequate quality of information available to decision-makers at the top of the state pyramid.

For the self-employed and cooperative–1 modes there may be another problem. They imply that each member invest both capital and labor in the firm. The same individual may, however, have different preferences regarding the use of labor and capital: a person may want to remain a member of a cooperative but may prefer to invest his capital elsewhere. This inability to separate the two factors also leads to inefficiency in the use of resources.

A capitalist joint-stock company will face the problem arising from the split of the entrepreneurial income. Whoever (management or shareholders) may at some point exercise the entrepreneurial role will not be the sole recipient of the entrepreneurial income.[4] In other words, the decision-makers will not be the only ones to profit (or lose) from the effects of their decision. The same wedge between the actual and appropriated yields exists in public corporations where the board of directors—even if alone responsible for management—will not be able to capture all of the entrepreneurial income.

Types of firms where we do not detect inefficiency in the sense of Statement 1 are the cooperative–2 type, the closed capitalist firm, and the entrepreneurial firm. All imply (1) private ownership of capital, (2) undivided responsibility for entrepreneurship and the appropriation of revenues derived therefrom, and (3) the separation of labor and capital. It is also interesting that the entrepreneurial role in one case belongs to labor, in the second to capital-owners, and in the third to an autonomous factor not linked with either capital or labor.

Let us now consider labor. It can be owned, as we saw in chapter 1, by (a) another individual or state, i.e., not by the worker himself

Table 3.1

Reasons for Inefficiency

Type of firm	Reasons for inefficiency
Privately owned capital	
Owner-worker	Inseparability of labor and capital (nondiversification of assets)
Cooperative–2	—
Capitalist (joint-stock)	Split of entrepreneurial income
Entrepreneurial	—
State-owned capital	
State-socialist	Information gap
	Severance of the link between decision making (on use of capital and management of the firm) and reward; weak link between worker's effort and wage
Labor-managed	Severance of the link between decision making (on use of capital) and reward
Public corporation	Severance of the link between decision making (on the use of capital) and reward; split of entrepreneurial income

(slavery), or (b) the worker himself. Form (a) is inefficient because the efficiency of the human agent increases if he is allowed freely to make decisions about his work. This stems from the same root as the "information *lacunae*" problem, and the severance of the link between decision-making and the appropriation of income: if a worker's effort and reward are unrelated, it is clear that productivity will suffer. Similarly, "grass-roots" knowledge about ways to improve production is available only to him, while the owner will (not unlike the central planner) lack information about possible opportunities. For all the other systems (viz., those with legal freedom of labor) we can assume that the effort and reward are broadly related. Obviously, the stronger the relationship, the higher the productivity of labor will be. If the relationship is relatively weak—as in the case of state-socialist and sometimes public and large joint-stock companies—productivity, and the incentive to improve it, will be low. Still, in these cases, with the possible exception of the state-socialist firm, there are no inherent systemic reasons why the linkage between the two may not be strengthened.

Table 3.1 summarizes the main causes for the inefficiency of different modes of production. The key reason for the economic inferiority of centralized as compared to decentralized forms of organization derives from the inability of the former to make full use of (dispersed) economic knowledge. They thus face an information gap. The key reason for the lesser efficiency of state ownership of capital than private ownership of capital lies in the absence of direct and personalized responsibility for the outcome of economic decisions regarding the use of assets—that is, in the severance of the link between the agent who makes the decision and his returns from that decision.[5]

In the framework set out here we can easily see why state intervention creates inefficiencies. By reducing the range of choice for individual decision-makers (interference (a)), it denies the owner some possible uses of the resource, and thus violates our condition in Statement 1 that the owner (within a given legal framework) solely and freely decide on the use of the resource. By rendering the individual decision-maker not entirely responsible for the results of his actions (interference (b)), it creates a wedge between the actual yield and the return received by the owner of the resource. (It is immaterial if the wedge is in favor of the owner; e.g., if he receives subsidies.) Finally, by imposing additional costs on decision-makers through taxes (interference (c)), the state again introduces a wedge between the actual and the appropriated returns.

Chapter 4

Ideological Grounds for Liberalization

"Unfortunately, efficiency is a fact and justice a slogan."

—Jacques Ellul*

4.1 Manifestation of a Changing Attitude toward the Role of the State

Reasons for the current movement toward liberalization of economic life reside in a disenchantment with welfare economics in the West and the failure of state socialism in the East. A combination of the two is responsible for the retrenchment of the state in the Third World.

The ascendance of the anti–welfare-state orientation of public opinion in the West is reflected in three developments: a recent increase in importance (in terms of electoral success) of conservative parties, a shift away from a pro–"big government" position by the socialist parties, and an almost complete "socialization" of the communist parties.

As evidence of the first element, the most frequently cited examples are those of the United States and the United Kingdom. The Reagan and Thatcher "revolutions" have, because of their simultaneity, the importance of the countries involved, and their relatively long duration, captured popular attention. The analysis of the longer-term electoral

---

*Jacques Ellul, *The Technological Society*, Vintage Books, New York, 1964, p. 282.

Figure 4.1. **Center–Right Strength in Four Major Countries***
(% of seats in parliament).

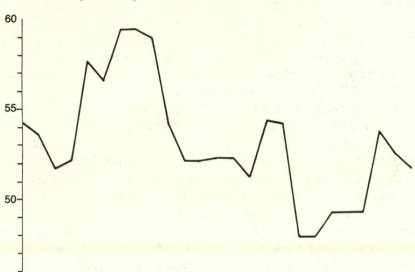

*United Kingdom, France, West Germany, Italy.

trends shows the remarkable stability of the conservative (center–right) parties, punctured with periodic spurts (and, of course, eclipses). Figure 4.1 shows the average electoral share of these parties in four major European countries (the United Kingdom, France, West Germany, and Italy). Their share—measured in terms of the percentage of seats held in the lower houses of national parliaments—ranges between 50 and 60 percent. There are three periods of electoral success of the conservative parties. The first extends over a relatively long stretch between 1968 and 1972, with student unrest and the oil crisis respectively at its beginning and end. The second period is 1978–1981, when the center–right parties in three countries (the United Kingdom, West Germany, and Italy) were gaining; the overwhelming defeat of the French right in 1981 is responsible for the brisk decline in that year's combined average electoral share shown in Figure 4.1. Finally, the third period starts in 1986, with the return of the right to power in France, and the 1987 electoral victories of the conservatives in the United Kingdom, Italy, and West Germany.

Although the analysis of electoral trends is indicative of the shift in public attitude toward the welfare state, its usefulness is somewhat limited. Analyses that try to detect changes in public opinion concerning a particular issue from electoral swings implicitly assume that the political positions of the parties have remained unchanged and that the issue in question was the decisive factor that swayed voters. More specifically, such analyses are based on the premise that the economic policies of the major parties with regard to state involvement have not evolved between, say, 1965 and 1985, and that the electorate made its decisions solely on the basis of the proposed economic policies. Both assumptions are generally incorrect. Political fortunes of different parties, particularly in the short run, are explained by a number of factors other than the advocated economic policies. Personal popularity of politicians, foreign policy developments, changes in neighboring countries, political or financial scandals, etc., all significantly influence electoral outcomes. But also, the positions of different parties concerning the economic role of the state have evolved. The changes in the economic philosophy of the parties that have traditionally been most closely associated with pro-statist policies may well prove to be more important for the long-term evolution of the role of the state in the economic sphere than the recent electoral victories of the conservative parties. In particular, we have in mind a shift away from the pro-public-sector policies of the socialist parties in Western Europe and the abandonment of Marxism and of its anti-market bias by the communist parties.

Changes in the economic policies proposed by the socialist and communist parties are important because they indicate that, on issues concerning the level of state interference in economic life, a broad consensus on a number of points may be emerging across the political spectrum. Although conservative parties may be more consistently anti-statist, socialist parties no longer view the public sector as the panacea for economic problems, as they did in the period after World War II. On the contrary, in many instances they increasingly see it as representing a drain on resources, a point of view that leads them to advocate a more limited role for the public sector as well as the transformation of state-owned firms into autonomous entities, in principle free from state interference. This last point constitutes liberalization, as defined in chapter 2.

The situation in France probably offers the best illustration of the trend. The Socialist victory in presidential and legislative elections led in 1982 to a number of nationalizations in banking and industry. The share of the public sector in total value-added jumped in that one year from around 13 percent to 16.5 percent.[1] The French industrial and financial public sectors, in relative terms, became the most important among all capitalist countries.[2] Yet, this dramatic expansion of the public sector did not lead to a marked increase of state intervention in the economy. Similarly, the independence of the previously state-owned firms was preserved. Indicative planning and the existence of "contract plans" provided a framework within which state ownership did not lead to centralized decision-making. In the newly nationalized industry, the entrepreneurial function in almost all cases rested with the firms' management, while the state simply replaced private owners of capital. In terms of Table 1.1, the developments in France could be represented as a move from the joint-stock company (cell(1,3)) to the public corporation (cell (4,3)). As explained above, such a change in the type of ownership by itself does not imply increased state interference.

Further evidence of the socialists' changing attitude toward state intervention is provided by contrasting the early post–World War II nationalizations in France and Great Britain with the French nationalization drive in 1982. Earlier nationalizations, as expressly stated in the preamble to the French Constitution of 1946, were motivated by the idea that "Any enterprise which has . . . acquired the characters of a national service or of a monopoly, should become the property of the community."[3] Note, in particular, the wide net that is cast by the concept of "national service," the equally ambiguous and euphemistic introduction of "the property of the community," and finally the use of the verb "should." The emergence of the public sector, as evidenced in the language used, was regarded as an integral part of a general strategy of reining-in of the market forces and displacement of the capitalist sector, at least from the areas where it was held to be particularly ill-suited to foster development. The public sector moreover was not considered in terms of its economic efficiency alone; i.e., simply as an alternative social arrangement. Quite to the contrary, the role assigned to it was much grander than purely economic. It was thought of as a single conflictless block, juxtaposed to the capitalist sector (with the motive of private profit supposedly replaced by "public interest"), and not as a conglomerate of individual and vastly different public firms with often mutually conflicting interests. The road travelled by European socialist parties since that time can be described as leading from a preference for the state sector,[4] with the state as entrepreneur, to the autonomous public corporation, which is fully part of the market.

By "socialization" of the communist parties in Western Europe we mean that, in economic matters, these parties have moved far enough toward the position of the socialist parties to be indistinguishable from them except in rhetoric. It is important to realize that communist parties no longer possess an alternative model for the organization of the economy. A comprehensive system of state planning, which they advocated after World War II and probably until the mid-1960s, was discredited by the performance of countries where it was implemented.[5] Even the support for a decentralized public sector has gradually been eroded, as its introduction in Western Europe has failed to live up to its promises. Communists have been obliged to abandon their positions one by one. They trail a few years behind the socialist parties in their evolution toward the center, and face the dilemma of losing their individuality (and getting assimilated by socialist and other left-wing parties) or sliding into marginality.[6]

In matters that interest us here, this means that the idea of a centralized, state-controlled economy has given way to an advocacy of indirect state involvement in economic life, more active redistributive policies, etc., all *within* the general framework of a capitalist market economy. Currently, communist parties in Western Europe (with the exception of the Portuguese, "external Greek," and factions of the Spanish and Finnish parties) no longer have an alternative model of society to propose.[*] To put it simply, the difference between socialist and communist parties after the war was that socialist parties advocated a major change *within* the capitalist system, which was supposed to take place by a gradual expansion of the public sector, while the communist parties advocated a radical break by pushing nationalization far enough to replace entirely the decentralized mode of organization (based on private ownership of capital) by a centrally planned economy. The failure of the Soviet experience has been such that communist parties cannot any longer propose it as a model to emulate. They have, of course, continued to concentrate on issues of income distribution. But it is important to realize that, according to any interpretation of Marx-

---

[*]In effect, no major Western European communist party can any longer be called Marxist, if we define the latter in terms of (a) introduction of a centrally planned state economy, and (b) espousal of the labor theory of value. (Obviously, other definitions of Marxism are possible, but they are irrelevant for our purpose because they do not concern strictly economic issues of the organization of production in a society; beliefs in historical materialism or the theory of alienation are not related to the role of the state in economic life.)

ism, income distribution is a derived problem (i.e., a problem derived from the private ownership of capital), so that concentration on such issues alone befits "reformist" socialist parties, but not the radical parties of social change.[7]

4.2 Central Planning and Development

We have seen how (i) the electoral successes of conservative parties, (ii) abandonment of the idea of centralized management of the public sector by socialist parties, and (iii) the loss of a credible blueprint for a radical economic transformation of society by communist parties, all illustrate a fundamental shift toward economic liberalization in the West. We shall now review the ideological elements that underpin this shift.

The first element has to do with a growing realization of practical problems associated with centralized management, the rising costs of the welfare state during times of economic recession, the budgetary pressures of inefficient state enterprises, and, finally, increasing evidence of superior performance by the more market-oriented segments of the economy. In light of these problems a more fundamental issue began to be glimpsed. As the West European economies rebuilt after the war, and then grew and integrated rapidly, the number of economic agents and, of course, of possible links among them, increased.[8] In general, as the economy becomes more complex, problems of information—already present at an early stage of development—become the true bottleneck. The nature of the crucial problem that an enterprise must solve, changes. In an underdeveloped economy, the number of possible links among economic agents is small because of (a) the importance of the subsistence economy (e.g., subsistence-producing peasants, who have no interaction with other farmers or city-dwellers); (b) the importance of traditional relations (sales or purchases made by or from the traditional traders from the same village or social group); and (c) undeveloped transportation, which in itself hinders greater integration of the economy. In such a nondiversified economy the essential problem is a technical one: how to produce a given good. Problems of demand, competition, etc. are subsidiary to the key problem of availability of resources and technical know-how. Capital and skilled labor are fundamental bottlenecks to production. In a diversified economy, both capital and skilled labor are available. The crucial economic problem becomes what to produce, whether to diversify or not, how to reach consumers, where to buy inputs, how to manage the financial portfolio. All of these are problems of information. Coordination becomes more complex as there is an increasing number of economic agents whose decisions are of relevance for an enterprise, and as consumers' preferences, due to greater amounts of discretionary income, become more diversified and more difficult to satisfy. Economic information and the knowledge of "place and circumstance" are thus of primary importance for the success or failure of an undertaking. The inadequacy of centralized economic management, which cannot access these bits and pieces of information, therefore appears more clearly than in a less developed and less integrated economy. Inefficiency can be seen to arise directly from the fundamental inability to use opportunities. Growth and diversification of European economies, as well as economic integration of the western half of the continent, have put in bold relief these information-related elements that make central planning difficult and inefficient. They were not so manifest in the period immediately following the war, when the task of reconstruction was a relatively straightforward one.[9]

In general, central planning is an efficient type of organization to realize tasks when objectives are relatively simple (nonconflicting) and information is technical in nature and can therefore (unlike economic information, which is dispersed) be centralized. This is best seen in the increases of production achieved through centralized planning during wars and, during peace-time, in rapid arms build-ups or recoveries from natural disasters.[10] The outstanding examples of the first are German planning in world wars I and II (the German experience in the first war having had a strong influence on the Soviet adoption of planning)[11] and British planning during World War II. Examples of the second are the German expansion of military production in the period 1933–39 and the Soviet military build-up after the last war. An example of the third is the rapid recovery of Bucharest after the 1977 earthquake (in a country that is otherwise noted for its dismal economic performance). The problem, however, is that such circumstances ("centralizable" information and straightforward objectives) are not normal circumstances of economic life.

The second reason lies in the changed nature of technological progress. The view that the state sector was the way of the future was also supported by technological determinism, which held that technical progress necessitated large-scale production. The latter, of course, could best be accomplished by the state thanks to its ability to mobilize resources. Just as we today think of electronics and computer industry as the most technologically advanced sectors of the economy, and most countries try to latch on to these developments, in the first half of the twentieth century it was electricity generation, gas and oil, steel production, and railways that were the gateposts on the road to progress. And, indeed, all of these undertakings required large savings often beyond the scope of individual capitalists, development of complicated networks, and some form of state compulsion to bring them about. With their large economies of scale, they were the antithesis of perfect competition. It is symptomatic that Hayek in 1944 felt compelled to write: "the myth is deliberately cultivated that we are embarking on the new course [of state planning] not out of free will but because competition is spontaneously eliminated by technological changes which we neither can reverse nor should wish to prevent" (1944, p. 43). Indeed that *was* the prevalent attitude at the time.

4.3 The Failure of State Socialism

The third reason for a shift away from statism lies in the failure of central planning in countries where it was implemented. The failure was political, social, and also economic. Absence of economic freedom and initiative, which is implied by the state ownership of capital and central planning, was seen to have led to even greater limits on political and ideological freedom. This manifest failure of the East European countries permeates the move toward liberalization in the capitalist economies, the reform attempts in state socialism, and reduction of the role of the state as an agent of economic transformation in the less developed countries.[12] The failure of the state-socialist model has become more and more apparent over the last fifteen to twenty years. The lag of East European behind West European countries was not as obvious in the mid-1960s as it is in the late 1980s. This was due to a combination of factors. First, growth rates in Eastern Europe were higher then than the growth rates in the West. It was their subsequent continuous decline, as well as closer attention to the actual meanings of these rates, that created the awareness of the relative decline of the socialist economies.[13] Second, the technological revolution focused attention on the inability of these countries to innovate and to introduce new technologies commercially, and often even to apply technologies developed in the West with an approximately equal level of efficiency.

This inability was all the more striking because socialist countries had, by any identifiable (objective) criterion, the capability to do so: they disposed of a skilled labor force, social peace reigned, bureaucracy was less venal or incompetent than in many other countries. Therefore something else which could not be encompassed by these indicators— the very nature of the system—must have been the cause.[14] Third, the failure was also concealed by a general perception, still prevalent in the sixties, that the then-existent difference in the standard of living between East and West was mostly the product of an inherited underdevelopment of Eastern Europe, predating the introduction of central planning. This view implicitly lumped all the East European countries together and failed to recognize that there was no development lag prior to the introduction of central planning for Czechoslovakia or East Germany. Even where relative underdevelopment did exist (as in the case of, say, Poland or Romania), it was with respect to Great Britain or France, but not to Greece or Spain. The awareness that a gap was emerging between centrally planned Eastern Europe and the Mediterranean countries was not present twenty years ago.

These factors explain why economists, and the general public as well, became increasingly disposed toward theories that explained the (now visible) gap between Eastern and Western Europe in terms of system-induced impediments to growth in the centrally planned economies, rather than differences in respective starting points. The emergence of the same realization among large segments of the intelligentsia and the general population in Eastern Europe, combined with a greater freedom to publish such views, and easier communication between the two parts of the continent, contributed to swaying public opinion in capitalist and less developed countries toward viewing central planning as essentially a failure, at least insofar as improvement in the population's standard of living and technological development were concerned. Finally, as a *coup de grâce*, the return to postwar-style rationing of food, energy, and a number of consumer goods in some socialist countries (Poland and Romania) in the 1980s—more than thirty years after such practices disappeared in Western Europe—highlighted the fact that something was seriously amiss with the system. It is inconceivable today to imagine, say, Spain imposing food rationing on the order of that introduced in Poland in 1981–82 (meat was limited to three kilos per person per month, and butter, sugar, rice, and flour put on ration cards). It is equally difficult to imagine Greece blocking out electricity to consumers for up to twelve hours a day, cutting television programming to two hours a day, and banning the use of private automobiles, as was done in Romania in 1984 and 1985.[15]

The general atmosphere of pessimism that has gripped Eastern Europe has a lot to do with the realization of socialism's failure. This mood can best be understood by again comparing the situation in the mid-1960s with that in the mid-1980s. When a number of state-socialist countries (Poland, the USSR, Hungary) embarked on reform projects in the mid and late sixties, economists and social scientists in these countries believed that reforms would help unleash the energy needed for development, and that the system would be adaptable enough to generate fast growth. The prevalent perception was that the remedy for the ills of the economy could successfully be found within the framework of the existing mode of production. In that period people like Leszek Kolakowski, Adam Michnik, Jacek Kuron, and Edward Lipiński in Poland, Ota Šik in Czechoslovakia, and Rudolf Bahro in East Germany were still ideologically part of the socialist movement.[16] It was failure in the realization of the stated (economic and social) objectives of the reform, and the gradual reassertion of both the immovable ideological character of state socialism and the role of bureaucracy, that transformed such people into opponents of the system. Few are ready to believe now that the link-up of socialist countries with the modern world of fast technological change can be accomplished by timid reforms, tinkerings with the type and the number of plan indicators, or experiments in selected branches of industry. On the contrary, the general perception is that the very political and economic set-up of the system impedes faster economic development, and that economic change is inconceivable without political reform that would radically alter the system as we know it.

The Gorbachev reform seems to hail back to the period of optimism, as yet another attempt to improve the economy without proceeding to substantial changes in the economic system. The Deng Xiaoping policy, on the other hand, seems to derive from the realization that the centrally planned framework itself needs to be revised in some of its essential components.

We shall discuss the prospects for liberalization of socialist economies more thoroughly in chapter 5. Here let us simply mention that throughout the period the problem of efficiency of central planning posed itself differently for the Soviet Union than for the other East European countries. The Soviet Union displays a number of characteristics (natural wealth; lower original level of development; docile population with a low standard of living, traumatized by the successive ordeals of revolution, civil war, Stalin's terror, and World War II; great-power status, and the evident success of the central planning system in enhancing the global power of the Soviet Union; fewer contacts with the West) that are not present in Eastern Europe, such that the economic inefficiency of central planning was felt less acutely in the Soviet Union than in the rest of the bloc.* The situation reversed in the 1980s, however, when rapid technological progress threatened to leave the Soviet Union militarily in a position of inferiority vis-à-vis the United States, and thus to weaken its global role politically and eco- nomically. The economic awakening of China, coupled with the Soviet fear of that country, further underscored the deterioration of the Soviet position. Economic immobilism thus threatened the status of the Soviet Union as a great power. Yet the same immobilism did not affect the position of other countries whose international role was, in any case, very limited, and by virtue of their size was bound to remain so under any economic system. There, the system was acceptable to the elites as long as it kept them in power and external stimuli to change were absent. This explains why revisions in the system of central planning are now more likely to come from the Soviet Union than from any East European country. To paraphrase Lenin, as far as the preservation of the existing model of state socialism is concerned, the Soviet Union is now the weakest link in the chain.

---

*Objectively, central planning was probably less inefficient in the Soviet Union than elsewhere, due to both the lower initial level of development and the exis- tence of vast untapped natural wealth. The last point underscores the fact that cen- tral planning is effective in *activating* resources; and so long as they are plenti- ful and cheap, the growth performance is relatively satisfactory. But the system is inefficient when it comes to economizing these resources when they cease to be abundant and cheap. These two facets are reflected in the ease with which the popu- lation was transferred from agriculture to industry and was able to find em- ployment there (e.g., in the USSR, Bulgaria), but also in unwise investments of cheap foreign credits in the 1970s, which led to the debt problem of the next decade. The system was thus able to suck unused resources (human or natural wealth) into the production machine fairly smoothly; but, lacking a mechanism to ensure their efficient use, it was incapable of making (what is called in the East- ern European economic terminology) the transition from "extensive" to "in- tensive" development.

Part II

THE POLITICS OF LIBERALIZATION

In the next two chapters we shall study reform in the context of state-socialist and capitalist economies. By reform we mean a movement that generally goes in the direction of liberalization. However, while reform reduces the role of the state in some areas, in others state intervention may increase. In order to keep terminological clarity we shall therefore reserve the term "liberalization" only for situations where there is effectively diminished state interference in the sense of points (a)–(c) in Definition 7. For the overall process we shall use the term "reform."

Chapter 5

Reform in State Socialism

"If we adopt the thesis that enterprises are independent . . . what scope is left for guidance and restrictions?"
—Miaczyslaw Rakowski*

5.1 The Economic Structure of State Socialism

In order to study liberalization in state-socialist countries in all its economic and political ramifications, we need to identify the social structure of the system, i.e., the different economic groups and their positions in the system. Only after the completion of this analysis can we proceed to a study of groups that are likely to favor liberalization and groups likely to oppose it, and to an analysis of the form that liberalization in state-socialist countries might take.

In state-socialist societies we find five modes of production. They are, in descending order of importance: state-socialist, owner-worker, labor-managed, public-corporation, and capitalist. As we indicated in chapter 1, the state-socialist type plays the predominant role. The small-scale proprietorship (owner-worker) type is essentially limited to agriculture and some services. This type is sometimes, as in Poland, Hungary, and China, very important in agriculture, but its significance outside agriculture is fairly limited in all socialist countries. The 1987

---

*A Politburo member of the Polish Communist party, in an interview in *Magyar Hirlap*, June 11, 1988, p. 1.

reforms in the Soviet Union allowed self-employment and the formation of owner-worker firms for the first time since the NEP (New Economic Policy) period of the 1920s.

The labor-managed sector also exists mostly in agriculture (cooperatives in Hungary). More recently in Hungary and in China it has been making inroads in manufacturing, trade, and services. In Hungary it functions as either a "franchise" or a "contract" system. Under the first, state-owned capital is leased to individuals who run the business as they wish, and pay the state a fixed rent. Assets are leased to the individual who offers to pay the highest rent. The system is prevalent in retail trade (11 percent of all shops in 1984) and the restaurant business (37 percent in 1984).[1] The "contract" system is somewhat different. Groups of workers in a state-owned firm hire (with the management's permission) the firm's equipment after regular working hours. They may produce something for the use of the firm itself, or may contract a job for an outside organization; the firm (and thus ultimately the state) receives a rental income. A similar situation exists in China, where smaller manufacturing enterprises are leased to workers.[2]

In several of the socialist countries most involved in the reform process, public corporations are becoming increasingly important. A significant portion of the enterprises in China, Poland, and Hungary stand somewhere mid-way between classical state-socialist firms and public corporations. A large percentage of their output is sold on the free market and most inputs are acquired there. Management follows only a few binding state targets or none at all. The legal status of these enterprises is still the same as that of state-socialist firms; yet, as they move ever closer to the pure type of public corporation, it is likely that they will also acquire a separate legal status.* This should help their expansion and independence, by delineating the limits of state interference in management more clearly than is the case today.

Capitalist enterprises are by and large concentrated in services (e.g., the restaurant business, crafts, repair shops), but their growth is circumscribed by the legal limits on the number of workers that a private employer can hire, which exist in most state-socialist countries.[3] This sector is thus severely constrained and its expansion is, for ideological and political reasons, viewed with suspicion.

Four of these five types of organization imply decentralized coordination of economic activity. Only the state-socialist mode implies cen-

---

*This is already happening in Poland where firms entirely owned by several state enterprises may acquire autonomous (limited liability) status.

Table 5.1

Appropriation of Income in State Socialism

Type of Firm	Type of income			
	Labor	Capital	Entrepreneurial	Taxes
State socialist	Workers	State	State	State
Owner-worker*	Workers	Workers	Workers	State
Labor-managed	Workers	State	Workers	State
Capitalist	Workers	Capitalists	Capitalists	State

*Includes self-employment and Cooperative–1.

tralization. But since it is the dominant mode, the economy is properly called state-socialist.

5.2 The Social Composition

In accordance with our analysis in chapter 1, the forms of appropriation of income will vary with different types of firms. Table 5.1 shows the appropriation of income by different social groups in state socialism.[4]

On the basis of Table 5.1 we can identify the following social groups: (1) workers in state-owned firms; (2) owner-workers (essentially farmers); (3) workers in capitalist firms; (4) workers in labor-managed firms; (5) capitalists; and (6) the state.

Obviously, there is no social group called the state. One must therefore identify the groups whose incomes are derived from state revenues. Here we can speak primarily of two social groups: (a) the political bureaucracy and (b) technical and managerial cadres. Political bureaucracy comprises all the state apparatus, including professional functionaries of political organizations (e.g., trade unions, the Communist party, various party-controlled organizations), and the army and the police. Technical and managerial cadres are responsible for the management of the state-socialist enterprises. Their income also derives from the overall state revenues, because they operate in a centralized structure in which their performance is assessed according to the degree of fulfillment of the plan targets or some other bureaucratic criterion, and does not stem from an independent (market) success or failure. Consequently, although their incomes often include bonuses, bonuses cannot be regarded as entrepreneurial income, since they are not earned independently by the managers, but represent transfer of a part of state revenues.

Among the political bureaucracy we can distinguish between high-level and middle-level bureaucracy. The high-level bureaucracy are the state and party professionals occupying high positions at the center. These would include permanent higher-level staff of the party central organs, the most influential members of the Central Committee, ministers and their aides in the central government, top military brass, and, obviously, the Politburo and the Council of Ministers (cabinet members). What differentiates high-level bureaucracy from middle-level bureaucracy is the scale at which it operates: the middle- and low-level bureaucrats operate at the commune, city, or regional level; high-level bureaucracy operates at the central level.* The middle-level bureaucrats may often be members of high party organs (e.g., the Central Committee), but they do not participate in the day-to-day management of political and economic affairs of the country as a whole.[5] Their remoteness from the effective levers of power imparts to them a different perspective on the system and its evolution. Their role, however, is not that of a simple transmission belt. While they stand below the central (high-level) bureaucrats in the hierarchy of power, and thus are in some sense subordinate to them, they also represent the constituency for the high-level bureaucracy. Their support is often indispensable in order to get certain policies accepted and carried out. Similarly, their passive resistance can abort policy changes. More ominously, their displeasure may find champions among higher-level bureaucrats, who can try to ride to power carried by the middle- and low-level bureaucratic reaction. Therefore, they may be as often ''courted'' by different factions at the top as they are ordered.

In some particular cases the top echelon of the military can be separated from the rest of the high-level bureaucracy and regarded as an independent group. The experience of state-socialist countries to date, however, does not generally warrant such a division. The military are

---

*The distinction between the central (federal) level and the regional level of bureaucracy, which exists in all state-socialist countries, is not a product of the heterogeneous ethnic composition of some of these countries (e.g., the Soviet Union, China, Czechoslovakia). The same division appears in nationally homogeneous countries (e.g., Poland), because it is dictated by the nature of the system itself. When the state controls the entirety of economic and political life, all of its activities cannot be carried out from only one place. Implementation of the decisions taken at the center requires some delegation of responsibilities, which leads to the formation of regional state and party organs. These organs partly follow the orders of the center, but, since they ''control'' the economic life in their regions, they also represent particular regional interests, which they try to press on the central bodies.

well blended into the state and party apparatus.

There are two more social groups whose incomes are derived from state revenues: the first are state administration employees (including workers in social services, e.g., health, education, etc.), and the second are recipients of transfer incomes (e.g., pensioners, invalids, students, people on welfare, etc.).

Figure 5.1 illustrates the social structure in the system of state socialism. All the identified social groups are shown in thickly drawn boxes. Direction of payments (from the economy to households, and from the economy to the state to households) is indicated by arrows.

It may be useful to note that, conceptually, income in all cases represents net income, i.e., income adjusted for different explicit and implicit subsidies. In real-world calculations, the impact of subsidies (e.g., controlled prices of essential goods, or a below-market level of home rents) on different social groups is difficult, but not impossible, to assess.

5.3 Liberalization: Reduction of the State Sector

If we inspect Table 1.1 and recall our definition of state intervention in the economic sphere, it becomes clear that pure state socialism (that is, when only the state-socialist mode of production exists in a country) is an example of complete dominance of economic relations by the state. In such a situation, the state would control almost all decisions of economic agents (our element (a) in Definition 7); the agents, not being independent, cannot be financially responsible for the few (if any) economic decisions they take (our element (b)); and the state would appropriate all nonlabor income, i.e., capital and entrepreneurial income, in addition to taxes (our element (c)). Obviously, such a situation represents only a theoretical possibility. Actual state-socialist systems show a blend of different modes of production, with state interference in the operations of other modes (owner-worker, capitalist, and labor-managed) being less extensive.

Because of the complete control exercised by the state over the state-socialist sector—a feature that is inherent to the system—liberalization in state socialism must always take the form of *abandonment* of the state-socialist type of production and a move toward the other types. While a reduction in the role of the state in capitalism (e.g., by lifting price controls) is possible within the *same* (capitalist) mode of production, every movement away from comprehensive planning represents a

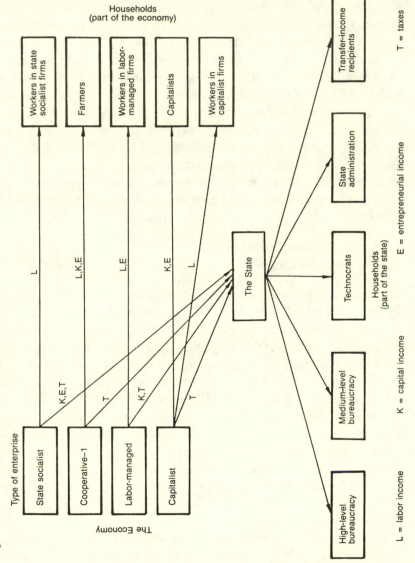

Figure 5.1. **Social Structure in State Socialism.**

movement toward a *different* type of organization. To illustrate: As the number of compulsory targets conveyed to an enterprise by the center decreases, the enterprise obtains discretionary power over a set of decisions. In these areas, horizontal links among enterprises begin to replace vertical links (i.e., central planning), and the coordination mechanism becomes decentralized. This is a slow process in which both types of links may coexist, rather uncomfortably, for a while. To use a term coined by Kornai, a command system of direct bureaucratic control is replaced by a system of indirect bureaucratic control. Nevertheless, indirect bureaucratic control is much more in line with the traditional role of the state in a decentralized economy. Thus the centrally planned system is being slowly eroded by the increasing importance of firm-to-firm links.

This movement is illustrated by the Chinese reform. The influence of state organs (at the central, provincial, or county level) over firms' decisions has radically diminished. Illustrations include the charging of interest rates on investment capital received from state banks (instead of pure grants as in the past) and the introduction of a corporate profit tax (instead of automatic full remittance of profits to the state). Both, according to our classification, represent a movement from the state-socialist to the public-corporation type. Although the power of local and regional authorities, growing at the expense of the center since the late 1960s, has not been sufficiently curbed, the changes nevertheless represent the extraction of the firms from under the state planning writ. [6] Similar systems have been dubbed ''bargaining'' in Hungary or ''folksy bargaining'' in Poland. ''Folksy'' comes from the fact that the firm does the bargaining with its ''own'' bureaucracy (middle-level regional bureaucracy), which on the one hand tries to control the firm as closely as possible, but on the other hand tries to shelter it from outside competition or interference of central organs. ''Folksy'' also because the rules are practically nonexistent: everything depends on the relative power of the agents and their bargaining skills.

Experiences in China, Poland, Hungary, and, most recently, in the Soviet Union show that the evolution away from the state-socialist system is carried by means of the so-called ''state orders.'' State orders are centrally determined quantities of goods that individual enterprises (at state-set prices) are obliged to deliver to the center. For these orders the state ensures the supply of required inputs. They are thus no different from the usual central planning, except that instead of covering the entire production of the enterprise, they relate only to a portion of its total (expected) output. In essence, they represent a way to scale down the extent of central planning. As the relative importance of state orders declines, the part of the economy where decentralized decisions determine the level of production expands. In 1987 in Poland, for example, the part of the economy that could be considered as centrally planned comprised less than one-half of total industrial production. This included the core of the economy, for which mandatory central material balances are drawn (e.g., iron and steel, liquid fuels, coal, grains, fertilizer, etc.; in total 114 products of "strategic" importance), plus export targets stemming from the government's bilateral trade commitments (mostly with other Soviet-bloc countries). Specific features of the "centrally balanced" goods are that (a) their production and export targets are broken down by enterprises, (b) they are compulsory (with the state ensuring the supply of inputs, inclusive of imports), and (c) their prices are determined by the state.[7] Point (b) obviously requires central planning not only of the "strategic" products (whose output is 100 percent planned), but also of a part of the production of other goods that are used as inputs. The share of centrally allocated inputs in the total consumption of intermediate goods provides a relatively reliable estimate of the extent of central planning in force (this because the control that the state retains is the most effective in the area of input supply). As Table 5.2 shows, the share of centrally allocated inputs has been more than halved in Poland during the last ten years.

A similar evolution is evident in Chinese industry, where dual systems of state orders and free market also coexist. In the first, the prices of inputs and outputs, as well as the supply of inputs, are guaranteed by the state; in the free-market part of the economy, prices of both inputs and outputs are generally higher and the firms' main problem is the difficulty in obtaining necessary intermediate goods. Market distribution, however, has been steadily rising. A survey of more than 400 large industrial enterprises shows that in 1985 market sales of output and market purchases of inputs amounted to respectively 44 percent of total output and 27 percent of total input supply (both up, from 32 and 16 percent respectively in 1984).[8]

In general, persistence of shortages and underdevelopment of the wholesale trading network are the prime obstacles to growth of the free market. Obviously, both are related to the existence of planning: state orders preempt most of the inputs, and wholesale trade will not develop if there is almost nothing to trade. We thus see that state orders, coupled with assured supplies of inputs, permit planners to

Table 5.2

Poland: Percentage of Centrally Allocated Inputs

1978	1982	1985	1986	1987
90	70	50	45	35

Source: World Bank (1987, p. 119).

maintain central control over the composition of production.

The compulsory nature of state orders at government-determined prices makes state orders different in kind from state purchases in decentralized systems. In the latter, the state appears merely as another consumer. In the centrally controlled part of socialist economies, on the other hand, the state treats enterprises the "old way"—as subdepartments of the national economy. It is the reduction of that centrally planned part of the economy that provides an indication of the extent of liberalization.

In effect, liberalization in state socialism involves two types of movements. The first concerns the state-owned sector of the economy, where, as we just saw, liberalization is translated as a movement toward the public corporation (or, in some cases, toward the labor-managed firm). State ownership over capital remains, while the hierarchical management of the economy is replaced, at least in part, by decentralized coordination. The state sector is therefore partially or totally (depending on the extent of liberalization) transformed into the public, or labor-managed, sector. This movement is illustrated in Table 5.3 by arrows A and B.

The interests of workers and managers in state-socialist enterprises differ. Workers tend to prefer a devolution in the direction of labor-management (arrow B) because their importance and power would thereby be increased. Through workers' councils they would exert control over production decisions and distribution of income; they would also receive entrepreneurial income. Furthermore, an element of populism ("we can all be managers") underlies the workers' preference. Managers and technocrats, on the other hand, would normally prefer devolution toward a public corporation (arrow A in Table 5.3), where the entrepreneurial function belongs to the management. Their attitude is motivated by skepticism about workers' ability to manage firms efficiently (e.g., a tendency to distribute high wages, at the

Table 5.3

Liberalization in State Socialism

		Labor	
		Entrepreneurial role	Hired out
Capital ownership Private:	Entrepreneurial role	Small-scale proprietorship C	Capitalist
	Hired out	Cooperative–2	Entrepreneurial firm
State:	Entrepreneurial role	n.e.	State-socialist
		B	↓ A
	Hired out	Labor-managed	Public corporation

expense of investment). The interests of managers and workers coincide only to the extent that they are both in favor of less state interference. They diverge as soon as the question of the type of organization to succeed state-socialist firms is raised. Each group, understandably, supports movement toward that type of organization in which the entrepreneurial function belongs to it.

Resentment against the state-socialist (hierarchical) type of organization is probably greater among managers than among workers. This is because managers, by their very position, are bound to feel more immediately the harmful effects of bureaucratic meddling in management. They are also a more articulate and educated group than workers, and therefore more likely to question the way things are run. On the other hand, workers who believe that they would fare worse in a public corporation than in the existing state-socialist firm[9] might create a strategic alliance against decentralization with the bureaucracy. An acceptable and certain wage-level, job security, and a relaxed work atmosphere could continue to be guaranteed to workers in exchange for their tacit acceptance of the system and, more specifically, of the bureaucracy's dominant role.[10] This alliance will be of particular importance at times when technocrats may be on the verge of winning their conflict with the bureaucracy. Workers who are not normally active participants in the political process can then be activated by the bureaucracy, to prevent or undermine the change. We shall explore this scenario more fully in the following sections; but it is important to note it here, in order to illustrate how the interests of workers and managers in the state sector, which at first sight seem to coincide, are effectively different, and how the interplay of such interests may thwart any movement toward liberalization.

5.4 Liberalization: Expansion of the Private Sector

The second liberalization movement involves that part of the economy in which productive assets are privately owned. We have seen that two such modes of production exist in state socialism (owner-worker and capitalist). Liberalization normally implies an increase in the importance of these two types. This is accomplished by loosening the state regulations that inhibit their growth. Owner-worker or capitalist firms may be allowed to enter a number of branches hitherto reserved for the state sector. Thus, for example, the 1987 Soviet law on individual labor opened up about forty types of activities, and the Soviet law on cooperatives (June 1988) allowed cooperatives (i.e., partnerships of at least three people with no hired labor) to operate in most industries and services. Private retail trade was permitted in China in 1982 and in Yugoslavia in 1983. In Czechoslovakia, all services, without restrictions, were opened to the private sector in 1988. In Poland, according to the new law passed at the end of 1988, the private sector can operate in practically all areas of activity.

Probably the greatest opportunity for the growth of owner-worker types of organization lies in liberalization of agriculture. The dominant form of organization in agriculture is/was state-socialist (kolkhozes and sovkhozes in the USSR, communes in China, cooperatives in Czechoslovakia) or labor-managed (cooperatives in Hungary).[11] As shown by the Chinese experience, the existing arrangements in agriculture could be changed easily by giving peasants land on a long-term lease (actually, a free lease, as in China) and then making it transferable.[12] The land thus only formally remains state property. Differently, land holdings may be almost entirely privatized, as in Poland. Unlike in industrial activities, where it is difficult to see the likelihood of a shift from the types of enterprises in which capital is state-owned to those in which it is privately owned, the prospects for privatization in agriculture (arrow C in Table 5.3) are much better. Reasons for this are the following: (a) the already established fact of greater economic efficiency of private as compared to state holdings,[13] (b) the peasants' strong preference for small-scale proprietorships over all other arrangements, (c) the relative ideological acceptability of privatization so long as it involves small-scale production with little or no hiring of outside labor, (d) the existence of similar arrangements in other state-socialist countries, and (e) most importantly, the fact that the peasantry was never regarded as the prime constituency of the "socialist transformation" and its political role has historically been a fairly passive one.

Point (a) is important not solely because of the usual efficiency considerations, but also because problems in agriculture leading to shortages in food products have greatly increased the political vulnerability of socialist countries to pressures from abroad. Since at least one part of the political establishment regards relations with the external world as fundamentally antagonistic, greater self-sufficiency will be a strong argument in favor of privatization of agriculture. Point (b), for example, is not present in manufacturing, where, as we saw above, workers seem to prefer transformation of state-socialist into labor-managed, and not privately owned, firms. The main reason for the workers' lack of interest in privatization, and the peasants' strong preference for it, resides in the difference in what privatization means in each case. Privatization in manufacturing does not change the position of the workers. If anything, the immediate effects, except for a minority of skilled workers, would be negative: possible loss of jobs, tighter work discipline, and little or no wage increase. Conversely, privatization in agriculture effectively hands a parcel of land back to the individual peasant. Clearly, if privatization in manufacturing could be conducted as in agriculture, so that the enterprise would in effect be sold to workers, their interest in privatization would be greater. Yet, because of the difference in financial strength, even an initial hand-out of shares to workers (or a sale at below-market prices[14]) would eventually result in standard "outside" privatization (i.e., shares owned by capitalists). Technological indivisibility of plants precludes the sort of "Lockean" transfer of ownership ("unification" of the worker and the object of his work) that is possible in agriculture. It is an undeniable fact that the link between an individual worker and his enterprise is not as strong as the link that exists between the peasant and his land.

Point (e), the fringe role of the peasantry, also helps its case for privatization. This happens in two ways. First, the doctrine of industrialization to which state-socialist societies without exception implicitly adhere assumes that the importance of the agricultural sector (particularly in terms of employment) declines. If so, it matters less which mode of production is prevalent in a sector that is bound to shrink. The matter is altogether different for the core sectors, which determine the character of the whole economy. (It is also important to realize that in all contemporary political systems the elite exerts its role in *cities*. See section 1.8, above.) Agriculture is therefore more apt to be liberalized without jolting the existing system of distribution of power. Second, Marxists have traditionally ascribed to peasants a preference for private ownership. Thus, if decaying classes, like the peasantry, express their preference for privatization, this falls within the "preordained" pattern of things. It does not threaten the essential tenets of the ideology. But, if the same preferences are expressed by workers, the very foundations of Marxism are threatened, for it is precisely in the name of the industrial proletariat and its assumed preference for "socialization of the means of production" that the revolution was carried out. The fact that peasants toil on their own private plots of land does not matter greatly to an average bureaucrat; but the fact that workers no longer need his "leadership" deprives him of political power and undermines the ideological justification for that power. Thus preference for privatization can ideologically be tolerated in the peasantry, but not in workers (where, as we mentioned, it is weaker anyway).

Finally, point (d), i.e., the fact that privately owned agriculture has remained dominant in some state-socialist countries (Poland), or was reintroduced (China) without this having had a detrimental effect on the power of the bureaucracy, is an additional element that should help privatization in agriculture.

Liberalization would also involve an increase in the importance of the capitalist type of organization. Its increase, however, is unlikely to be as significant as the expansion of self-employment and owner-worker firms. This is, first, because of the lesser ideological acceptability of capitalist firms, which in socialist societies implies their political unacceptability; and second, because unlike in agriculture, there is no clear constituency to lobby for the reintroduction of capitalist firms.

The first point, translated into practical terms, means that constraints on the growth of the capitalist sector may at best be only marginally relaxed. For example, the ceiling on the number of privately hired employees may be raised, more activities may be thrown open to the capitalist sector, or joint ventures between capitalist and state-owned firms may be allowed.[15] However, all this in itself, as long as the official climate remains fundamentally hostile to private enterprise, is unlikely to produce any dramatic change. The essential obstacles (high tax rates, difficult and long licensing procedures, legal uncertainty, a great number of regulations to be followed in daily affairs and their random enforcement) will most probably remain. And easily the most formidable obstacle, though impossible to quantify, is the one that has to do with "atmospherics." As long as the official ideology regards private enterprise as an alien element, a kind of *modus vivendi* between the bureaucracy and capitalists can be established only if bureaucrats treat the private sector as a milking cow that, for its own "protection," can be spoiled of a fair share of its profits. They are helped in this by a widespread egalitarian ideology and by people's abhorrence—inculcated by the ideology—of those who stand out. Successful capitalists could easily be accused of profiteering, speculation, etc., and their wealth seized. Their private wealth stands out in sharp contrast to the low salaries of those employed by the state sector, and, unlike in less developed capitalist countries, they lack the political clout to fight back. The question will invariably be asked: What kind of socialism is it, in which state-sector workers are poor and wily entrepreneurs are rich?[16] Of course, opening the doors of the private sector wider would reduce such anomalous profits (results of the absence of competition), but would it not also undermine the power of the bureaucracy whose natural "field of activity" is the state sector, and in the long run threaten the system?

The capitalist sector thus begins to resemble those semilegal activities which are tolerated by the authorities (as long as they pay well for "protection") but may at any instant be banned. Not surprisingly, in such a hostile environment, the essential prerequisites for a successful capitalist development (security of property, reinvestments of profits, and a long-term view) will be absent; the sector will operate at some low level of activity. In order not to attract unwanted attention, the most successful entrepreneurs will not only avoid expanding their activity, but will stop short of maximizing even the profits from current operations. They will become masters in dissimulating their success. In Eastern Europe one can often hear complaints voiced by private businessmen, not only that they could employ additional workers, but that they feel compelled to stop work at some percentage of the *allowed* capacity—in order not to be too conspicuous.*

The second point mentioned above, namely the lack of a natural constituency for the introduction of a private sector, is obvious. The only people with a strong and immediate interest in real liberalization

---

*A different reaction, stemming from the same cause, is characteristic of those with extremely short time horizons: maximize current profits to the utmost, do not invest, and move out of the business quickly. It will be noted that both types of behavior are similar to those in illegal activities: in one case a thief tries to camouflage himself by living modestly, in the other he undertakes a few daring forays and then quits.

of the capitalist sector are either people who are presently part of that sector (i.e., very few) or those who believe they would do much better in a liberalized capitalist sector than they are now doing elsewhere. These are similarly few, as the capitalist-*cum*-entrepreneurial talent is not very abundant in any society, and is not likely to have been helped by a historical record of stifling private initiative. Furthermore, it is not entirely evident that even those who are currently in the capitalist sector would welcome liberalization of that sector. While liberalization would eliminate some constraints on their activity, it would also leave the field open to new competitors. Since most of the existing capitalists may be appropriating high rents due to their monopoly position, liberalization, greater transparency of the rules, and so forth hold significant risks for them.[17] We are thus witnessing at least tactical complicity between capitalists and local bureaucracy. The coincidence of interests is not stable and can, at any instant, be broken. Nevertheless, in state-socialist societies it introduces a feature that did not exist under the conditions of a quasi-complete absence of a private sector, and that one is more wont to find in capitalist countries with strong government interventionism (and relatively corrupt bureaucracy). But this collusion between capitalists and a section of bureaucracy further narrows the active constituency supporting liberalization of the capitalist sector.

We have so far considered only small-scale capitalist firms such as they exist now in socialist countries. What are the prospects for private investments in enterprises, either in the form of bonds or equity? On the surface, the problem is ideological, but more fundamentally it has to do with the redistribution of economic and political power. The reintroduction of private property over assets for people who are entirely unconnected with the process of production clearly goes against the fundamentals of Marxism.[18] The issue, however, differs for bonds as opposed to equities. In the case of bonds, investors appear merely as lenders who do not aspire to a role in the management of the firm; their claim on income derives only from the property of capital. Although ideological objections apply, in principle they are no greater than objections to any private appropriation of income on the basis of property. Bonds, as we shall see, have little effect on the *modus operandi* of the economy and the distribution of economic power. Indeed, private bonds for the financing of state enterprises have been used in several socialist countries, and bond markets exist in Hungary and China. The one in Hungary has been in operation since 1983 and bond holdings of the population amount to some 5 percent of its savings deposits. Although this is a fairly minimal amount,* private capital invested in a state-owned firm means that the return from the ownership of capital is now shared between capitalists and the state. The entrepreneurial function, which, as we saw, is the crucial test for the classification of the modes of production, remains in the hands of the state. The firm is thus still of the state-socialist type. The introduction of private capital therefore does not have any practical impact on the position of the firm, and in particular on the prevalence of the vertical vs. horizontal links with the rest of the economy. Similarly, the position of a private joint-stock company in capitalism is not altered in any fundamental way if the state has some participation in it. Introduction of bonds (normally guaranteed by the state) thus expands investment opportunities for private individuals and allows state firms to raise some financing;[19] it does not have an impact on essential features of the system.

Introduction of equities would make a difference. Since the return on equity is linked with the profit of the firm, private investments will be forthcoming only if investors (1) have a say in management and (2) can be sure of the accuracy of the financial results of the firm (i.e., that the profit is truthfully reported). The first point immediately implies a change in the nature of the firm, since the entrepreneurial function would need to be taken away from the state and handed over to private investors. The firm would therefore cease to be a state-socialist firm. The second point implies the introduction of strict and enforced accounting rules. In their absence the equity financing easily translates into depriving investors of any return by simply declaring zero profits. Since private investors are not exactly the class of people whom the state is most keen to protect, they would have little recourse.

---

*It could be argued that, at the same level of development, financial markets would necessarily be less developed in socialism than in capitalism because of less sharp wealth (as opposed to income) differences. Some groups (e.g., the high-level bureaucracy) may dispose of extremely high incomes, particularly if implicit revenues (e.g., below-market rents, access to special shops, free holidays, etc.) are included and assessed at their scarcity prices. But none of these perquisites allows one to build wealth. Thus the approximately same level of income inequality in capitalist and socialist societies, reported by most studies, translates into less *wealth* inequality in socialism. It is not an accident that socialist societies allow for high income differences, but not for accumulation of wealth. Egalitarianism is not the primary reason for it. More important is the fact that wealth can be the basis for an individual's independence from the authorities. The very high level of consumption provided to the high-level bureaucracy, combined with the inability to accumulate wealth, ensures obedience to the state: the rewards of obedience are clear, and there is no independent wealth to fall back on.

The realization of either of the two points (but particularly (1)) in any sizeable portion of the economy would, in effect, result in the abandonment of the state-socialist mode of production and of the whole organization of society and distribution of power that goes with it. It is therefore no surprise that so far nothing resembling this has happened in socialist countries.[20] In addition, as noted above, there are no identifiable social groups that would push for this solution (in effect, a return to capitalism). One can therefore expect that, if equity markets were to appear, they would be very limited in size. They would be so, not only because the state wants to keep a lid on them, but also because there would be few investors unless the conditions cited above were fulfilled; and it is difficult to see how, short of a revolution, they could be. Their hypothetical introduction could represent only a tactical, and hence transitory, concession by the hard-pressed bureaucratic leadership.

In conclusion, liberalization in state socialism involves a change in the structure of ownership. Liberalization is two-pronged: it proceeds by reducing the size of the hierarchical state sector in favor of the public or the labor-managed sector, and by increasing the importance of the sectors where capital is privately owned, in particular owner-worker firms and (non–joint-stock) capitalist firms.

Let us consider now what happens to the role of the state in economic life. A movement away from the centralized system implies by definition a reduction in the level of state intervention. It must accordingly enlarge the economic agents' sphere of free decision-making. Divestiture of the state sector means less determination of economic parameters by the state and greater accountability of economic agents for their own decisions (our definitions (a) and (b) of state interference; see section 2.1, above). The level of taxation must also go down because, in a classic state-socialist system, firms remit all of their profits to the state.[21] For the first time (as can be seen, for example, in the Soviet law on state enterprises) firms dispose of some discretionary income, which—still following the central guidelines—they are able to distribute among investments, wage increases, and social benefits for workers.

The increase in the relative importance of the private sector must, almost without exception, be accompanied by reduced state intervention in that sector. Examples include raising the legal maximum of workers that a private firm can employ, expanding the areas in which the private sector can operate, and so forth. As for the (b) type of state interference (accountability of economic agents), one component of it

Table 5.4

Reform in State Socialism

Change	Liberalization		
	(a)	(b)	(c)
Reduction of the state sector	−	−	−
Increase of the private sector	− (?)	−	?

Notes: (a) = state determination of economic parameters.
(b) = subsidies and protection.
(c) = taxation.
− denotes less state interference (liberalization).
? indicates uncertainty about the direction of change.

(subsidization) is nonexistent in socialism anyway, since the state does not bail out or subsidize private firms; the second component (shelter from competition) is reduced by the expansion of the private sector itself, because liberalizing the licensing policy and letting newcomers in exposes the existing firms to increased competition. Then, unless there is concurrently an increase in other forms of state intervention such as price fixing or higher taxation,[22] expansion of the private sector should indeed represent liberalization in the exact sense as defined in chapter 2.

Table 5.4 illustrates what can be expected concerning different aspects of liberalization (our points (a)–(c)) following a partial divestiture of the state sector and an increased role for the private sector. A minus sign denotes reduction of state interference (i.e., liberalization); a question mark indicates uncertainty about the likely direction of change.

The degree of liberalization of the economy can be examined by looking at the structure of production. Table 5.5 shows that in Hungary in 1984, almost 35 percent of national income (two-thirds of it accounted for by labor-managed firms) was produced outside the state sector; that proportion increases to 85 percent in agriculture. In China similarly, the level of liberalization is far greater in agriculture than in the national economy as a whole: the (hierarchical) state sector accounts for almost three-quarters of industrial output, while it is practically totally absent in agriculture. Data for Hungary and China are particularly revealing when contrasted with some other state-socialist countries (Czechoslovakia, East Germany, the USSR) where, as we saw in section 1.9, the importance of the state sector approaches 100 percent.[23]

Table 5.5

China and Hungary: The Structure of Production (in percent)

	Hungary 1	Hungary 2	China 1	China 2
(1) State sector	65.2	15.3	73.6	1.8
(2) Labor-managed sector	20.6	51.1	0	0
(3) Public corporation	0	0	10.8	0
*State-owned capital**	*85.8*	*66.4*	*84.4*	*1.8*
(4) Owner-worker	8.7	33.6	14.2	98.2
(5) Capitalist sector	5.5	0	1.4	0
Privately owned capital	*14.2*	*33.6*	*15.6*	*98.2*
Decentralized coordination				
(2) + (3) + (4) + (5)	*34.8*	*84.7*	*26.4*	*98.2*

*Production in enterprises with state ownership of capital.
Notes:
  Hungary 1: In terms of national income in 1984.
Labor-managed sector = cooperatives (land is not private property of which peasants can dispose). Owner-workers = household farming + auxiliary production of employees. *Source:* Kornai (1986, p. 1692).
  Hungary 2: In terms of gross agricultural output in 1984.
Kornai (1986, p. 1701). Same definitions as above.
  China 1: In terms of gross industrial output in 1984.
Collective-owned industry (at provincial and city level) most closely approximates the public corporation in view of the greater influence exerted by managers, their more immediate access to relevant bureaucrats, and the fact that production is outside the central plan. All of rural industry (village- and township-run) classified as cooperative-1 (owner-worker). Capitalist sector = local private sector + joint ventures. *Source: Statistical Yearbook of China 1986,* pp. 182, 227.
  China 2: In terms of agricultural households in 1984. *Source:* Sah (1986, p. 57).
Owner-workers = households under the "responsibility" system.

A similar analysis of the level of decentralized coordination in the economy can be made from Figure 5.2, which shows the percentage of Polish national income produced outside the state sector. It ranges between 15 and 20 percent, with the notable exception of 1981, when disruptions in production, strikes, etc., cut the output produced by the centrally planned part of the economy by 16 percent, while the private sector grew by 7 percent and its share in national income accordingly increased to almost 30 percent. Before that, massive investments and expansion of production during the Gierek period (1971–80) steadily eroded the private sector's relative importance; the same appears to be happening in the recent period, despite all the official pronouncements

Figure 5.2. **Poland: Share of Non-State Sector in National Income (%).**

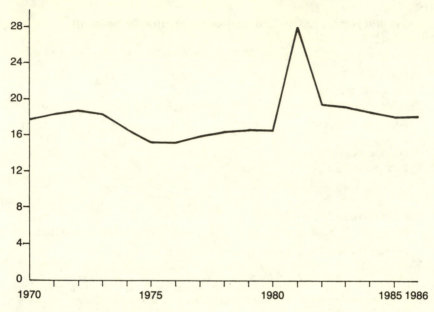

Shares calculated at current prices

Source: Rocznik Statystyczny, various issues.

in favor of the private sector. However, since a significant proportion of state enterprises are evolving toward the public-corporation type, and the centrally planned core of the economy is shrinking, a simple contrast between the (officially defined) state sector and the private sector underestimates the extent of decentralized coordination currently existing in the system.

5.5 The Central Conflict in Socialism

The decision to reform the economy is always a political decision. It is particularly so in state socialism, which is a political society *par excellence*—that is, a society in which politics, and not economics, rules. We must, therefore, first identify the main political actors and their interests. We define as main political actors those who are autonomous, meaning those who directly participate in the political process and directly influence political decisions. In the next stage we shall look at other social groups that are important but are not autonomous actors;

they are "harnessed" as allies by the pro- or anti-reform forces. They may impose constraints over one type of development or another, but they are not direct participants in the political process. To make the distinction clear, we can say that the bureaucracy is an autonomous social group; workers are not, although their interests (and strategy) will have an (indirect) influence on the outcome of the political process.

In section 5.2 we identified ten social groups in state socialism. Only three of them represent autonomous political actors: high- and middle-level bureaucracy and technocrats. These groups directly determine the direction of policy because they participate in its shaping.

The central conflict in state socialism takes place between middle-level bureaucracy and technocrats.

We have defined as technocrats all people who are in positions of influence in firms (e.g., managers, economists, engineers). While economists and engineers are technocrats par excellence, managers need not be so. Among managers there are also middle-level bureaucrats who, while pursuing a political career, are temporarily filling managerial positions. Such managers are essentially bureaucrats in disguise, and they will have more in common with the ordinary middle-level bureaucracy than with technocrats. However, a majority of managers will be people who, either by training or preference, have a technocratic outlook. By "technocratic outlook" we mean that they are concerned with efficiency as such. They are people who want a well-run enterprise, good-quality products, and tight work discipline.[24]

Technocrats resent the vast middle-level bureaucratic meddling. Since enterprises lack autonomy, and cannot rely on market relations to ensure smooth functioning of the firm, technocrats find themselves at the constant mercy of middle-level bureaucracy. It is the bureaucrats who, through their influence, help the firm get credit, scarce inputs, or foreign exchange. How do bureaucrats get into a position where they wield such influence? Here we must rely on the seminal work of Janos Kornai, *Economics of Shortage*.[25] Explained in simple terms, the mechanism by which this happens is as follows. The origin of the "shortage economy," which characterizes all state-socialist countries, lies in the existence of soft budget constraint. Soft budget constraint means that firms are not responsible for their financial results. They know that their losses will ultimately be covered by the state: the firms cannot go bankrupt.* Soft budget constraint will understandably make firms insensitive to costs. Since their "field of action" is not limited by the budget, they will exhibit the "growth syndrome" ("investomania," as it is also called). The firms will try to produce and invest as much as possible, regardless of costs and efficiency. They will be helped in this by the presumption that derives from the central-planning philosophy, namely that the *volume* of output, the *volume* of investments, etc. are what really matters. (This, of course, makes perfect sense within the terms of the system: if prices in a centrally planned economy are meaningless, the only unambiguous indicator of success is physical size. It is therefore not an accident that all the state-socialist countries put an emphasis on physical indicators.) To grow, the firms need inputs (labor, capital, foreign exchange). As they are not too concerned with their prices, firms will try to buy as many inputs as they can. The same type of action from all the firms in the economy will produce a general shortage. The firms will then hoard inputs, making the shortage worse. At that point, (middle-level) bureaucracy must step in to allocate the shortage. Why bureaucracy? Because it must be done by a group that is "above the fray" (managers clearly cannot fight it out among themselves) and has the political power to impose decisions. Bureaucracy fits that role perfectly. Bargaining thus remains the only allocational mechanism. Like all bargaining, its outcome will depend on the relative power of the participants (i.e., bureaucrats). This is why it will be important for managers to have powerful bureaucrats to whom they can turn when production threatens to stop, ordered inputs fail to arrive, or banks refuse credit.[26]

A relationship of symbiosis between technocrats and middle-level bureaucracy does not normally develop, however. As a *quid pro quo* for

---

*The question may be asked: Why would the state always stand behind a firm? There are a number of reasons. First, firms in a centrally planned economy are not independent economic agents. They are subdepartments of a single whole (the national economy). The money earned by one firm will be transferred to another that is operating at a loss, just as within a single firm there will always be cross-subsidization. Second, letting some firms close down will create problems of unemployment, social conflicts, etc. Third, and most important, even if the state dedecided to close down "bad" firms, it does not know which firms are really bad. If a firm makes losses, this may be because it is obliged to trade at unrealistic prices; or the state might have forced it to produce what it normally would not. Without market prices to convey information about the relative success of firms, all decisions concerning closings will be equally arbitrary. So, not only will no single firm go out of business, but the structure of production will not be changed, since it is not clear that an alternative structure would be more efficient given the existing prices.

helping technocrats run the enterprise, the middle-level bureaucrats constantly meddle in the management. They prefer to keep technocrats on a "short leash" in a subordinate position of supplicant, and will constantly be on the lookout for "technocratic tendencies." These will include, for example, technocrats' stabs at greater independence, their concern with operational efficiency to the detriment of ideology, the willingness to innovate and explore new procedures, and so forth. The justification that middle-level bureaucracy offers for raising "the specter of technocracy" belongs to the realm of ideology: technocrats disregard the social and class content of production, subordinate it to the purely technical aspect, and thus come close to their capitalist counterparts. The position of a worker in such a firm—the argument goes—does not differ from that in a capitalist twin, and the party must, as "the vanguard of the working class," and through the intermediary of the middle-level bureaucracy, reassert the class aspect. Of course, ideological discourse is only a cover for the power struggle between the middle-level bureaucracy and the technocrats. But in this ideological struggle technocrats are singularly at a disadvantage. What bureaucrats are saying is indeed the official ideology of the state. They are also its certified custodians. Their access to the media is easy. Technocrats are not generally people very skillful in ideological arguments (which means that they cannot win by "packaging" their views into the official jargon), nor can they afford to explicitly oppose the official doctrine. Technocrats are thus on the defensive.

The power struggle aspect becomes very clear if we consider what would happen if the technocrats got their way. The elaborate rationale for the existence of middle-level bureaucracy would collapse. If the class element (the presumed guardian of which is bureaucracy) no longer matters, efficiency of production becomes paramount. And efficiency is delivered by technocrats. If enterprises become really autonomous, and technocrats solely responsible for their management, the links among the firms will become horizontal, a market will be introduced, and prices will begin to allocate inputs. What, then, is the function of the bureaucracy? Bureaucrats cannot run the enterprises, workers are no longer to be protected against exploitation, while bargaining and the need for rationing disappear. Conscious of that fundamental conflict, middle-level bureaucracy would ideally like to dispense with the technocrats altogether. This, however, cannot be done without severely damaging the economy, and it is difficult (although not impossible, as the Cultural Revolution and the Kampuchean experience have shown) to accomplish on a larger scale. On a smaller scale, however, it is being done continuously. If a head of an enterprise becomes too independent or successful, if the state and party bureaucrats in the region believe that the economic costs of firing him are sustainable,[27] they will do it. But on a larger scale it is difficult: so technocrats and the middle-level bureaucracy will remain locked in an uneasy relationship and will continue to vie for power. Economic difficulties will strengthen technocrats. Bureaucracy will then need technocrats to survive. But when the economy improves, bureaucracy will mount the counter-attack, and the balance will shift in its favor.

To a large extent political considerations explain both the current indebtedness of socialist countries and the urge for detente in the 1970s. The bureaucracy, in effect, had to face the fact that satisfying the rising aspirations of the population, partly attributable to the demonstration effect of Western Europe, required a faster increase in the standard of living. The question was how to deliver this standard of living without giving more power to technocrats. The answer was to fall back on capitalists: to borrow money from the West, introduce new technologies, and achieve growth through greatly increased capital infusions, without changing the economic system and disturbing the internal balance of power. As we know, the plan did not work: loans have to be repaid, and the grafting of the newest technology on unchanged relations of production gave only scant results.

The optimal situation from the bureaucracy's point of view is the one in which technocrats are at its permanent mercy. Mushrooming of regulations, licenses, laws, rules, decrees (often mutually contradictory or unenforceable) does not solely represent justification for the existence of bureaucracy (who else is to produce these laws and to oversee their application?), but, from a strictly legal point of view, is also intended to make any manager liable to dismissal at any time. Operations of the firms would virtually come to a standstill if all the regulations were strictly observed. Since managers cannot let this happen, they may often be encouraged by ''their'' (regional or communal) bureaucrats to disregard formal procedures. This in itself increases the influence of the middle-level bureaucracy: it decides which rules can be broken. Yet when the same bureaucrats want to dismiss the manager, a specific legal offense can always be invoked (e.g., land or water was acquired without a license, raw materials were bought on the black market, a special regulation on fire-emergency standards or workers' hygiene was not observed, etc.). It is, of course, immaterial if everyone else did the same, or if the bureaucrats originally agreed to the "offense."

In addition, proliferation of regulations may also be used to check uncontrolled growth and economic success of a firm. Success, if not associated with a particular politician, provides an independent power base for technocracy and is, by implication, a threat to the power of the bureaucracy. Any activity, in particular a successful one, undertaken without the express involvement of bureaucracy, is viewed by it as an implicit denial of its indispensability. As in the Middle Ages, when each human activity had to reflect religion, so each activity in socialist countries must reinforce the feeling that no success would have been possible were it not for the involvement of the bureaucracy. It is not failure that threatens the state apparatus; it is success to which it contributed nothing.

To sum up: A planned economy, with soft budget constraint on the enterprise level, and resulting shortages, is a precondition for the power of bureaucrats. Managers are further kept in check by proliferation of regulations, some of which are normally unenforceable without damaging production. Yet nonobservance of these regulations provides the legal basis for the dismissal of any manager. Discretionary application of sanctions consolidates the power of bureaucracy. To rule bureaucrats need, in short: (a) financial indiscipline, (b) proliferation of all types of regulations, many of which are impossible to implement or must positively be broken, and (c) discretion in the application of the law. Their power is strong only if the law is not obeyed, all economic agents are in constant contravention, and the punishments are discretionary.[28]

5.6 The Politics of Reform

Technocrats resent middle-level bureaucracy because of the system-induced dependence on them. Technocrats realize that if they were free to act as they saw fit, if horizontal lines of coordination (the market, for short) among the firms existed, they could ensure steady supply of inputs, reduce costs, etc. Their position allows them to see advantages of the market mechanism, and simultaneously disadvantages of rationing, price controls, or hidden unemployment. They see in their everyday activity that economic knowledge is indeed a specific knowledge of opportunities of time and place. Central planning for them translates into operations under constant stress conditions: they have to realize targets even if raw materials are not delivered on time, absenteeism is high, or machinery breaks down. A rapid recourse to alternative inputs (including labor), available in a market system, is excluded here because of widespread shortages and hoarding by all actors. Managers cannot punish or reward suppliers or workers with money; they can only implore: workers to work harder, suppliers to deliver the promised inputs, bureaucrats to use their political connections. It is a frustrating position for highly qualified and motivated people.

Thus the key conflict arises at the middle level of the hierarchy where the two well-defined groups have opposite interests. Technocrats will argue for the elimination of centralized allocation of inputs and output targets, for greater decentralization, and generally for an extension of market coordination. They will press for change by emphasizing efficiency. Middle-level bureaucrats will defend preservation of the *status quo* by stressing social costs of a move to a more competitive regime (unemployment, social segmentation), and by arguing that the introduction of decentralization and the market means creeping capitalism. Behind these arguments used in public debate are interests. Technocrats stand to gain with economic reform; their power, once they are free from the political meddling, will increase. Middle-level bureaucracy stands to lose: its power in the economic sphere will all but disappear. As efficiency, and not ideology, takes the commanding position, the basis for bureaucracy's social power is eroded. It is no longer clear what the function of the bureaucracy is: increasingly it becomes superfluous.

The most important factor in deciding the outcome of the conflict is the stance adopted by high-level bureaucracy. The high-level bureaucracy is the third and most important autonomous actor. In the conflict that opposes technocrats and the middle-level bureaucracy, the interests of the high-level bureaucracy are not as clearly defined as the interests of the other two groups. The essential interest of the high-level bureaucracy is preservation of power. This preservation depends, on one hand, on the level of support it receives from its constituency (that is, middle-level bureaucracy), but also on the economic viability of the system and hence its acceptance by other social groups outside the three main actors. The latter objective (economic progress to ensure social stability) only technocrats can deliver. Thence the truly ambivalent position of the high-level bureaucracy, which also explains why it cannot consistently support one group or the other. By its ideological preference it is led to rely on the middle-level bureaucracy; the economic realities force it to support technocrats. When the latter, emboldened by the reform, show signs of becoming too powerful, and thus in the long run of posing a threat to high-level bureaucracy itself, high-level bureaucracy will turn to its constituency and enlist its support in the struggle against technocrats. Forever divided between its preferences and economic imperatives, high-level bureaucracy will preside over a situation of permanent instability.

If high-level bureaucracy decides to move the economy toward reform, it will face two major obstacles. First, its constituency—the middle-level bureaucracy—will show its discontent, try to stall the reform, and will look for a group on top to champion its cause. The reformers among high-level bureaucrats thus put their own careers at risk. Second, high-level bureaucracy is aware that introduction of the market poses a long-term threat to its own legitimacy. The threat is not immediate as it is for the middle-level bureaucracy. Yet the same issue of superfluity, which instantly arises in the case of the middle-level bureaucracy, will eventually arise for the top levels.

There is, however, one reason why high-level bureaucrats will be more open to reform than their middle-level counterparts. It is that, dealing with macro issues, high-level officials are more conscious of the fact that tolerating increasing inefficiencies in the economic sphere also represents a long-term threat to their power. Economic problems may lead to popular discontent, strikes, military intervention, and a chain of events, the outcome of which, particularly for each individual politician, cannot be predicted. High-level bureaucrats thus face a cruel dilemma: even if their present position appears secure, both the reform *and* the continuation of the *status quo*, in the longer run, carry a possible threat to their power.

In section 5.3 we saw that workers in the state sector would generally favor a movement toward the labor-managed system, i.e., a different type of reform than advocated by technocrats. Thus, although they are not ardent supporters of the hierarchical state-socialist mode of production, they are also not allies of technocrats. They are more likely to prefer the current situation to the one where they would face stricter labor discipline, redundancies, and sharper wage differentiation combined with a probable (at least in the short run) decline in real wages.[29] Introduction of the market economy, accompanied by inflation and unemployment, would destroy the assured minimum that most workers enjoy without *immediately* providing any of the benefits, either in the form of direct participation in management or a tangible increase in real wages. In the conflict between middle-level bureaucracy and technocrats, workers in the state sector may therefore be expected to take the same position as the bureaucracy. Their influence on policy, however, is generally minimal until that moment when the middle-level bureaucracy begins to feel that the top echelons of bureaucracy may (for example, under the impact of growing economic problems) consider a move toward reform. When high-level bureaucracy begins to weigh an alliance with technocrats, thus directly undermining the position of the middle-level bureaucracy, workers will come to play an important role. They will be brought into the political arena by the middle-level bureaucracy, and their resistance to reform (with specters of impending social disorders) will be played on in order to dissuade top echelons of bureaucracy from an alliance with technocrats. The objective pursued by the middle-level bureaucracy is to persuade high-level bureaucrats that the reform will not be a quick-fix in terms of economic results, and may even lead to greater social strains than the current situation. In addition, the obvious link between economic and political freedom will be stressed, with the argument that the thaw (*"relâchement"*) cannot be contained in the economic arena, and that a loss of power for the top-level bureaucracy, instead of being averted (which is the very reason why high-level bureaucrats considered the idea of a reform in the first place), may even be precipitated. Tocqueville's observation that the most dangerous moment for a bad government is when it tries to become better finds its full justification there.

The support given by the workers in the state sector to anti-reformists may, in such a constellation of forces, prove decisive, and influence the entire course of events. The situation, however, is very precarious. If the middle-level bureaucracy and workers are strong enough to dissuade high-level bureaucracy from reform, the reform movement, defended by technocrats alone, is doomed. If, on the other hand, workers fail to give strong support to the middle-level bureaucracy, it alone has little chance of winning over an alliance of convenience between technocrats and high-level bureaucrats. Workers become the swing actor.[30]

Of the other social groups we identified, only peasants will not be a negligible group in support of reform. This, however, applies only if agriculture is collectivized. Peasants, as we have shown above, would then demand a return to the private ownership of land, and would support the reform movement. Their influence is not negligible because of the significance of agriculture for the sustenance of the social order. However, if agriculture is already mostly in private hands, the issue of the battle in the cities will be of little immediate concern to peasants, who are traditionally fairly removed from political centers of power.[31]

We have so far discussed the position of the five most influential social groups (high-level bureaucracy, middle-level bureaucracy, technocrats, workers in the state sector, and peasants). We shall briefly describe the position of the remaining five groups we identified in state socialism.

Transfer-income recipients (pensioners, students, invalids, the unemployed, housewives with children) will strongly support anti-reform forces; they are afraid that reform will mean inflation, in which case most of those living on fixed incomes are bound to be worse off. The relative position of transfer-income recipients is certainly worse in a market than in a planned economy.[32] The attention of people is often focused on their relative, not absolute, position, which, particularly in a crisis situation, is not an unreasonable attitude, since the two are then the same—the level of national income being assumed to be fixed.*

Small-scale property owners (artisans, small traders, restaurateurs, lawyers; i.e., the so-called "individual sector") and capitalists will display an ambivalent attitude toward the reform movement. Their attitude will depend on the level of constraints placed on their activity by the authorities and the existence or lack of a symbiotic relationship with local administration. If constraints on operations of the *existing* firms are relatively lax and control over the size of the private sector is exercised primarily through the licensing policy, the private sector may be lukewarm toward reform; for those who are already in, reform implies increased size of the sector and greater competition. Similarly, if the private sector and local bureaucracy have developed a symbiotic relationship in which the bureaucracy, through bribes (or even through extortion), receives a portion of the rent earned by private owners, the

---

*Even in times of prosperity the hypothesis that people may think in relative terms should not be discarded. Growth has a perceptible impact on the standard of living only when it is sustained for at least several years. A one-shot 2 or 3 percent increase is not felt. Cumulative increases, of course, require time. And since most people's time-horizons are short, a heavy discount applies to the longer-term gains. It is indeed an arduous task to convince people of the virtues of a policy by pointing to their long-term gains.

These points highlight the need to make a methodological assumption, when studying the effects of a particular economic policy in terms of winners and losers, to concentrate on the relative income of different classes. It is also justified by the fact that (in addition to income) *distribution* of power is a strong determinant of one's attitude toward a specific policy.

latter could be less inclined to upset a mutually acceptable balance.* More often, however, the situation will be such that private-sector firms are fairly closely controlled (e.g., in terms of employment and output), taxation is high, legal security is precarious, and the relationship between the private sector and the local administration is tense. Under these conditions, private owners are likely to support the reform because the field of permitted activities will expand (e.g., retail trade or construction may become open to the private sector), taxation could become less severe (a consequence of the official policy of private-sector encouragement), and, most important, legal protection of the private business may be reinforced. As mentioned above, the main problem private-sector people face in socialist economies is insecurity, the feeling that they represent an alien body in the system and that their activities could at any moment be discontinued, and even property nationalized. This problem dominates all others. A reform that promotes the idea of compatibility between the private sector and socialism, and provides a stronger legal guarantee for the private sector and property in general, will therefore be welcomed by small-scale property holders and capitalists. It is important to keep in mind, however, that both groups are fairly negligible in numerical terms,

---

*The level of symbiosis will depend on three factors: (a) the level of net profits realized by the private sector, (b) salaries of the local administration, and (c) its ideological commitment. The higher the (a), and the lower (b) and (c), the more likely will be the symbiosis between the private sector and local bureaucracy, and the existence of bribery. An example is provided by contrasting Poland and Yugoslavia in the 1970s and the 1980s. Local bureaucracy was better paid and more ideologically minded in Yugoslavia than in Poland; the actual level of control over the private sector was less, and the extent of bribery greater in Poland. It may be noted finally that element (a) depends on the restrictiveness of the licensing policy, fiscal policy, and the ease with which the private and state sectors do business with one another. In effect, the more restrictive the entry policy and the more liberal the fiscal policy, the greater net profits will be, and the greater the likelihood of collusion between the private sector and local bureaucracy. Also, if the private sector is allowed to deal easily with state firms, there would normally be a transfer of resources from the state-owned to the private firms (due to the existence of the soft budget constraint in state firms, their insensitivity to costs as well as outright corruption), and consequently an increase in private-sector profits. This is a phenomenon well known throughout Eastern Europe, so much so that state enterprises are either banned or discouraged from trading with private firms. We may thus find local bureaucracy supporting a stricter licensing policy toward the private sector simultaneously with reduction in taxation and a less restrictive policy toward dealings between the private and state sectors (for the existing firms). Interests of private owners and local bureaucracy thus coincide against the interest of the state as a whole.

Table 5.7

Tally of Anti- and Pro-reform Social Forces

Anti-reform	Pro-reform	Neutral
Middle-level bureaucracy	Technocrats	Workers in capitalist firms
Transfer-income recipients	Peasants	State administration
High-level bureaucracy	Capitalists	
State-sector workers	Owner-workers	

and even more so in terms of political power.[33]

Finally, two social groups are likely to adopt a neutral stance. These are workers in capitalist firms and the state administration. Their position will not, in any perceptible way, change under a more liberal economic regime. In the case of reform, administration will continue to implement decisions very much as before, and the presence of a number of people in state administration whose outlook closely resembles that of technocrats will be sufficient to offset those who might think that some loss of the state regulatory power will also affect their position as administrators. Table 5.7 shows a tally of anti- and pro-reform social groups. They are displayed in descending order by their commitment to one position or the other. For peasants this holds with the *caveat* noted above, i.e., in the table we assume that peasants work on state farms; with private property of land they might move to a neutral position.

5.7 The Calculus of Reform

We have identified ten social groups. We are concerned here with two attributes of each group—its political and economic power, and its degree of cohesion. If the group is completely cohesive we may say that we have broken the society into precisely relevant components for the study of a particular problem. We have seen, however, that for a number of social groups their level of cohesion, reflected in the attitude toward reform, is less than complete.

The first attribute is the amount of power to affect policies in the desired direction. If for a moment we assume that the level of cohesion is unity in each of our ten groups, then the tally of pro- and anti-reform forces from Table 5.7 can be translated in terms of power

Table 5.8

The Calculus of Reform I

Anti-reform		Pro-reform		Neutral	
Middle-level bureaucracy	9	Technocrats	8	Workers in capitalist firms	1
		Peasants	6		
Transfer-income recipients	4	Capitalists	2	State administration	5
		Owner-workers ("individual sector")	3		
High-level bureaucracy	10				
State-sector workers	7				
Totals	*30*		*19*		6

as in Table 5.8. There we give 10 to the group with the highest degree of social power (high-level bureaucracy), and successively decrease it by 1 until we reach the least influential social group (workers in capitalist firms). We shall assume that the ordinal scale is also cardinal, so that the "points" can be added.

Anti-reform forces seem much stronger. The pro-reform forces can win *only* if they attract high-level bureaucracy or state-sector workers. (Winning the middle-level bureaucracy over is, of course, out of the question.)

We shall now try to refine somewhat our analysis by introducing an estimate for the relative cohesiveness of different social groups. Obviously, the number of possible combinations for the victory of one coalition or the other increases as our social groups become less homogeneous. An illustrative example is given in Table 5.9. We assume that seven groups are homogeneous. Thus, for example, all middle-level bureaucrats and all technocrats are considered to be, respectively, anti- and pro-reform. Social groups that are assumed to be split are high-level bureaucracy, workers in the state sector, and the state administration.

What emerges from this analysis is that the two crucial groups for the success of reform are high-level bureaucracy and/or workers in the state sector. From the illustrative arithmetics in Table 5.9 it follows that either the majority of state-sector workers have to support the reform movement or the high-level bureaucracy must be split. These two conditions, which we established at the very beginning of our discussion, seem, regardless of the illustrative numbers in Tables 5.8 and 5.9,

Table 5.9

The Calculus of Reform II

| | Distribution | | |
Weights	Anti-reform	Pro-reform	Neutral
10 High-level bureaucracy	0.8	0.2	
9 Middle-level bureaucracy	1.0		
8 Technocrats		1.0	
7 Workers in state sector	0.7	0.3	
6 Peasants		1.0	
5 State administration	0.2	0.1	0.7
4 Transfer-income recipients	1.0		
3 Individual sector		1.0	
2 Capitalists		1.0	
1 Workers in capitalist firms			1.0
Weighted totals	*26.9*	*23.6*	*4.5*

to be essential conditions for the success of the reform movement.[34]

The crucial assumption, more important than the one about the homogeneity of different groups, is that about the relative power of the groups. We have interpreted ordinal ranking as reflecting equal cardinal differences in the level of power. This is probably acceptable in a situation of social peace, when changes in economic policy are piecemeal and represent the outcome of a balancing of the interests of various social groups. The distribution of power is obviously very different when open military force is used, or in conditions of social revolution. Thus, for example, in Poland during the "Solidarity" period workers ceased to be a nonautonomous social group and became a major actor. In the longer term, however, short of a major revolution that would radically alter or overthrow state socialism, the underlying structure of power outlined here will tend to reimpose itself.

We must thus distinguish between the core of a system with its attendant distribution of power among classes, and transitory alterations (principally due to social disturbances) of that pattern of distribution. The pattern derives from the fundamental characteristics of the system; the change represents the system's response to an emergency. If fundamental relations come out of the emergency unscathed, the original pattern of distribution of power will, with very few changes, reassert itself. This, of course, is true for any system: no less for

France before and after 1968, than for Poland before and after 1981. Although the May events and "Solidarity" wrought some longer-term social changes, neither of them altered the essential features of the society: France is still a capitalist, and Poland a socialist economy. These special times when the "normal" structure of power is disturbed (or, if the revolution turns out to be successful, a new "normal" structure is not yet established) represent a situation of "social disequilibrium." One such disequilibrium situation is when the high-level bureaucracy and the military (who are assumed to coincide) defend the system by force. The power of all other social groups is significantly reduced, and the power of some that are particularly heavily oppressed (e.g., peasants during Soviet collectivization) may fall to almost zero.

5.8 Applications: Poland, China, the Soviet Union

We shall apply the methodology developed so far to three recent instances of reform or attempted reform in actual socialist societies. We shall consider Poland in the period 1980–83, China in the period 1976–83, and the Soviet Union in the period after 1985. As we shall see, only in the first case (Poland) can we speak of a realignment of power during the most dramatic period of reform. In the two other cases, notwithstanding the importance of changes that occurred in China, the reform did not at any time affect the basic distribution of power in state socialism.

In Poland, the events of August 1980 which led to the birth of "Solidarity" represented a revolutionary change. The working class (state-sector workers), normally a nonautonomous social group, came to the forefront of the political battle. They became the main interlocutor of the high-level bureaucracy.[35] Most of the workers came out in favor of a reform that would transform state enterprises into labor-managed ones. This is the tendency that, as we indicated, could be expected to be stronger among workers of the state sector than among other social classes. In Poland, due to the sudden increase in the power of the (organized) workers in the state sector, and to the history of attempts to introduce labor-management (going back to the Poznan riots in 1956), this option—usually existent only in an incipient form—became realistic. The revolution that occurred in 1980 was brought about by a coalition of state-sector workers, technocrats, and, more hesitantly, peasants. The fact that workers and technocrats were (at least initially) united in an effort to overthrow state socialism explains why the forces of reform were at one point powerful enough effectively to threaten the basic structure of the system. The origin of the rather lukewarm support by peasants is to be found in the fact that peasants own their land and are thus, as pointed out in previous sections, more likely to remain on the sidelines of the main conflict. An obvious implication is that, had the land been state-owned (or effectively state-owned as in the kolkhoz type of cooperatives), the eventual outcome of the reform movement might have been different. The state-socialist mode of production might have been replaced in most of the economy by the public corporation, labor-managed, or capitalist types of organization, or, alternatively, the repression to which the state and the military finally had to resort might have been much more brutal. In such case direct Soviet military intervention in the face of the crumbling of the system in Poland, with all its unforeseeable consequences, cannot be excluded.

The system weathered the crisis more easily than expected, principally due to the existence of private ownership of land, which diffused the peasants' radicalism. It should be noted that the stabilizing role of the peasantry in this context is not new: as Marx noted in his "Eighteenth Brumaire of Louis Napoleon," the peasantry is generally a conservative social force. The irony of history is that these very characteristics might have saved a system based on Marx's ideas.[36]

The high- and middle-level bureaucracy, in most cases, belonged to the same anti-reformist camp. The explanation of the fact that high-level bureaucracy was less divided on the issue than we would normally anticipate is due to the unexpected and sudden nature of the revolution. In effect, reform movements in state socialism normally occur with at least the tacit approval of the high-level bureaucracy (or some part of it), which tries to improve the economic situation by a calibrated process of reform. It is thus normally at the helm of the process, and can (as we shall see from the example of China) control it. In Poland, however, the outburst was a grass-roots phenomenon that took the high-level bureaucracy by surprise. The movement was much more radical than the bureaucracy could tolerate: a direct and immediate conflict with bureaucracy was therefore inevitable. Instead of representing only a longer-term threat to the power of high-level bureaucracy, the reform movement clashed with it at the same time as with the middle-level bureaucracy (the latter, of course, was inevitable under any scenario).

It must be pointed out that it is the sudden reversal in the relative importance of different social groups, i.e., the emergence of the working class as an active force, that explains the term "revolution." The new arrangement was incompatible with essential features of state socialism. Either reform forces were to score a victory, and the system was to evolve away from state socialism, or the normal social structure of socialism had to be reestablished. The untenability of the situation means that it could not be other than a temporary situation between the two social equilibria—that is precisely what a "revolutionary period" is. When the high- and middle-level bureaucracy organized themselves for a decisive strike, they were able, by using the military, to alter the balance of force again in their favor, and to re-create the normal structure of state socialism.

Two conclusions can be drawn from the Polish experience. First, the working class historically appears to be unable to *lead* a successful movement for the overthrow of any system. This inability of the working class may have to do with the fact that any social system is hierarchical. Amorphous egalitarianism of the workers can produce a revolution, but it has no program to offer for the new system. Workers' revolutions are either put down (as happened in the Paris Commune, or in Germany, Austria, and Hungary in 1918–1920), or "hijacked" by bureaucracy, as happened in Soviet Russia.[37] The failure of Sorelian syndicalist schemes, as well as Lenin's realization that the overthrow of capitalism had to be directed by professional revolutionaries (i.e., by a party, and not by a social movement), a strategy that was crowned with success, are testimony to the desultory nature of exclusively workers' movements. The failure of Polish workers hails therefore from the same origin as failures of similar movements in nineteenth-century Europe.

Second, some aspects of the reform as defined by workers were not popular with other pro-reform groups. We have already indicated that workers and technocrats propound different reform directions (the first favoring labor-management and the second the public-corporation type of organization). The fact that the reform, because of the importance of organized workers in the movement, moved toward a labor-managed dissolution of the state sector, cost it support from technocrats and managers. Moreover, "Solidarity" itself did not develop a consistent program of reforms. While one section of workers supported devolution toward labor-management, the other expressed demands that were essentially syndicalist in nature and, more importantly, were inimical to the introduction of the market.[38] Views of the pro–labor-management section were compatible with the position of technocrats as far as introduction of the market was concerned; they were at odds with respect to the role of workers in the process of production. Exactly the opposite was true for views held by the "syndicalist" faction of "Solidarity." A syndicalist approach, which implies a sharp division of labor between workers and management, is consistent with the position of technocrats; but the rejection of the market is not. Moreover, the syndicalist faction of "Solidarity" opened the breach through which bureaucracy could satisfy the demands of the workers while leaving the system practically intact. The forces of reform were thus split, as was becoming more and more evident in the months leading to the December coup.

Having analyzed the failed attempt at reform through revolution in Poland, we shall now briefly review the reform movement in China. The most important difference is that the reform movement in China occurred within the normal framework of state socialism. The relative importance of different social groups was not disturbed as abruptly as it was in Poland. The reform movement in China stems from an alliance of the technocracy with a part of high-level bureaucracy. Throughout, it was directed from above. Unlike in Poland, where workers became an autonomous actor, state-sector workers in China are confined to their usual role of a nonautonomous social group. One of the ways for the reform forces to prevail lies in technocrats' succeeding in supplanting middle-level bureaucracy as the natural constituency of high-level bureaucracy, and thus isolating the anti–reform-minded "apparat." This was, it seems, realized in China. In addition, the reform was helped by the dissatisfaction of peasants who were organized in communes. Absence of peasants' ownership of land, their lack of freedom to decide what to produce and to earn money in accordance with their work, was responsible for making peasants a strong pro-reform force. Due to its sheer size and historical role in the revolution, the importance of the peasantry in China is probably greater than in other state-socialist countries. Deng, who allowed peasants to regain a *de facto* right to private property over land, essentially replicated Mao's pro-peasant policy of the 1940s which laid the groundwork for a Communist victory in the civil war. A remarkably rapid increase in agricultural production reinforced the position of the pro-reform forces (the pay-offs of the reform were visible to all) and firmly entrenched the peasantry in the pro-reform camp.

The original impetus for reform came from one section of the high-level bureaucracy. It was aware that it could count on the support of technocrats: the campaign to introduce technocrats (e.g., people who have made their careers as managers of state enterprises) into the state and party organs of power seems to attest to that. High-level bureaucracy tried thereby to change the nature of its own constituency. It could not trust the middle-level bureaucrats (who attained their positions during the extremist left-wing period of the Cultural Revolution by climbing ladders within the party apparatus) to implement a program of radical economic reform. The alliance between high-level bureaucracy and technocrats was based on a common objective—displacement of the middle-level bureaucracy. Whether this alliance can endure—in other words, whether economic reform can be realized without a fundamental change in the distribution of political power—is an open, and crucial, question for the fate of reforms, to which we shall return in chapter 7.

The only strongly anti-reform forces in China are still the middle-level bureaucrats, and some of the workers and high-level bureaucrats. The latter, as mentioned before, hope to ride to power carried by an eventual middle-level bureaucratic backlash. By the very nature of the system as it exists today, and the direction where it is heading, middle-level bureaucracy must remain inimical to reform. Peasants are virtually outside its writ; state enterprises, by becoming more autonomous, run by professional managers, increasingly extract themselves from under bureaucratic tutelage. A clear indication of the middle-level loss of power are the reports about local party and government officials who have become famous (and sometimes notorious) for engaging in assorted (licit or illicit) economic or financial undertakings. In order to remain economically on top they must swim with the current: and the current is such that now economy rules politics, and not *vice versa*. But clearly, even if some middle-level bureaucrats may thus enrich themselves, this is not a normal role for the social group as a whole. The basic enmity of middle-level bureaucrats toward reform, and their remaining power, are illustrated by the replacement of Hu Yaobang as the head of the party and the brief campaign against "bourgeois liberalism." In section 5.7 we saw, in effect, that reform may succeed if high-level bureaucracy is split. This has happened in China. Yet when the demands for reform go so far as to threaten the power of the high-level bureaucracy itself (as during the *dazibao* campaign in 1979 and the student demonstrations in 1985, for example), the alliance between the middle-level "apparat" and the high-level bureaucracy tends to reestablish itself. Demands for the "fifth modernization" (i.e., political reform) clearly affected the position of both. And when both wings of bureaucracy combine, the success of the reform is unlikely.

The crucial problem of state socialism, namely how to combine political and economic liberalization, is well illustrated by the Chinese example. As long as liberalization is merely economic (movement from the state to the public sector), technocracy can count on the support of one fraction of the high-level bureaucracy. Once reform extends to the political arena, the coalition of high- and middle-level bureaucracy is sufficiently strong to block it. Depending on circumstances, there may then be a backlash in the economic sphere. However, the likelihood of such a backlash in China remains weak at present, for the following reasons: (a) a relatively strong alliance between technocrats and one part of the high-level bureaucracy, (b) the tremendous economic success of the reform, and (c) the fact that the whole movement was conducted within the normal framework of state socialism without significantly altering the equilibrium social structure. The last point means that the "undoing" of the reform would probably also have to be done through the same channels. This would take a lot of time, consensus-seeking, etc., and make the whole enterprise subject to easy derailment. Point (b) is particularly important for China, as it is for the Soviet Union, because of the superpower status of both. As discussed in section 4.3, economic success is less important for preservation in power of the leaderships of small socialist countries, which obviously cannot have world-wide aspirations.

We shall briefly consider the likely evolution, in terms of the struggle between pro- and anti-reform forces, in the Soviet Union. The situation there, under the Gorbachev leadership, displays some, although limited, resemblance to the Chinese situation. The initiated reforms (e.g., greater independence of enterprises) have clearly upset the Brezhnevian relation of power between middle-level bureaucracy and technocrats, in favor of the latter.

As a digression we may note that in the two decades of Brezhnev's rule, the state-socialist system reached its theoretical ideal type. It established dominance of the middle-level bureaucracy over technocrats, and of politics over all other social spheres, and enabled the state to pursue its main objectives—military power abroad and security at home—to the utmost degree and with a high level of consistency. This was done while ensuring a necessary minimum of economic progress. Economic welfare thus clearly appeared as the constraint on the achievement of state objectives, and not as an objective in itself. It should also be noted that the ideal type of state socialism requires weak power of the top individual. This is necessary because of (a) totalitarian power vested in the state, and (b) the need for stability of the system. In effect, the totalitarian nature of the state combined with the dictatorial power of one individual leads, as Stalin's and Mao's cases show, to ultimate instability of the regime. The dictator's decisions spare no one, including the highest echelons of bureaucracy, from arbitrary treatment. They undermine the central pillar of the system and create an incipient instability.[39] A collective leadership of top bureaucracy lends security to the rest of bureaucracy and, by the very conservative nature common to all collective leaderships, guarantees the system against sharp policy zig-zags.

It is not clear that the Soviet reforms will emulate the reforms in China to their full extent. There may be two reasons for doubt. First, instead of decentralization and movement from the state-socialist type of enterprise to the public-corporation type, the projected Soviet reform seems to retain essential characteristics of centralized coordination. Greater freedom for individual agents will stem mostly from less detailed central planning and a gradually decreasing proportion of "state orders," so that the domain for individual enterprise decision-making—our point (a) in the definition of liberalization—will expand. As long as compulsory state orders do not begin to play a subsidiary role in the economy, most of the coordination of economic decisions will still be done at the central level and the system will not be changed. If, on the other hand, the importance of state orders diminishes in accordance with reformers' projections (to only 40 percent of total industrial output in 1990), it is not clear that the new arrangement would be viable. Every decrease in centralized decision-making introduces elements that tend to move the system even further away from the state-socialist type. Independent horizontal links among enterprises spring up. There is, in a sense, a tendency to overshoot the target level of decentralization. This, then, undermines the role of middle-level bureaucracy, and the show-down between it and technocracy becomes unavoidable. As we have seen, this conflict was solved in China in favor of pro-reform forces. In the Soviet Union, unless the power base (constituency) of high-level bureaucracy is changed, every gain in decentralization will be more difficult to sustain. In most general terms, this is, of course, the problem of whether economic decentralization can be compatible with the existing distribution of political power in the country.

The second difference with China is that the movement from the kolkhoz type of organization in agriculture to private land ownership seems likely to proceed at a much slower pace. Despite recent assurances from Gorbachev that land leases will be extended for up to 50 years, such a decision, in order to be successful, needs to overcome stiff resistance from the bureaucratic apparatus that has been built around state-owned agriculture and that would have to be dismantled, as well as a rather hesitant response it has so far received from the peasants. Second, the very excesses of the Cultural Revolution, and the fact that peasants have much greater political clout in China than in the USSR, made it easier to reverse the policy of communes in China.* In the Soviet Union, where collectivization happened more than a half-century ago, the weight of the *status quo* is much greater.

The outcome of the Soviet reform will depend on political factors: either the constituency will be changed from the middle-level bureaucracy to technocrats, in which case the reform could proceed further, or, if this power shift is not effected, the reform will follow its predecessors in leaving the essentials of the system intact. In the latter case, stagnation of the Soviet Union will continue. Awareness of this fact, in addition to the status of the Soviet Union as a major world power—which, if it is to be maintained, requires technological and military developments inseparable from a more efficient economic system—may strengthen the hand of technocrats. In such case an alliance of technocrats with a fraction of high-level bureaucracy, and in particular the military, cannot be easily dismissed. How far such an alliance could go, given the noncoinciding long-term objectives of the two groups, is the key issue.

---

*The extremist nature and the accompanying chaos of the Cultural Revolution have eased a later transition to a more liberal economy, first, by showing how left-wing extremism was unworkable and, second, by failing—precisely because of the amount of turmoil produced in the system—to create a stable middle-level bureaucracy interested in the preservation of the system. It may be noted that this bureaucratic interest in stabilization was singled out by Trotsky as the main reason for Stalin's success in the struggle against the proponents of "permanent revolution." The position of bureaucracy was much safer with "socialism in one country." Inability to create a strong central pillar of middle-level bureaucracy in China during the Cultural Revolution is also related to the exercise of despotic power by one individual. By analogy, we could say that the most propitious period for reform in the Soviet Union was just after Stalin's death, when somewhat similar conditions existed as in China after Mao.

Chapter 6

Reform in Developed Capitalism

"Our aim is to build upon our property-owning democracy and to establish a people's capital market, to bring capitalism to the place of work, to the high street, and even to the home."

—John Moore*

6.1 The Economic Structure of Developed Capitalism

In developed capitalist societies we find five modes of production; listed in approximate order of importance, they are: capitalist, public corporation, state-socialist, owner-worker, and entrepreneurial. All of them, except state-socialist, imply decentralized coordination. We shall briefly review these five modes.

The capitalist mode, as indicated in chapter 1, plays the dominant role. It can be organized as either a joint-stock or a limited-liability company. The joint-stock type historically followed the small-scale proprietorships using no hired labor[1] and the limited-liability company. Joint-stock companies are characterized by dissociation between ownership of capital and management. This is a feature that does not exist in other types of capitalist enterprise. It means that the entrepreneurial income is no longer entirely appropriated by capitalists, but may often be *de facto* shared between them and managers. Joint-stock firms are

*Financial secretary to the British Treasury (1984). Quoted in "Privatisation: Everybody's Doing it, Differently," *The Economist*, December 21, 1986, p. 85.

the dominant type of organization in developed capitalism. In effect their absence or presence is one of the key differences between mature (developed) capitalism and early capitalism in underdeveloped countries.[2]

Due to the importance of the state in developed capitalism, autonomous public companies probably represent the second largest type of organization. They are numerous in industry, but their principal areas of activity are energy and infrastructure, telecommunications, and transport.[3] As we saw in chapter 1, the entrepreneurial role in autonomous public companies belongs to management. In some public companies whose degree of autonomy is less, the entrepreneurial role may be shared by management, the state, and private shareholders (to the extent that they exist).

The less profitable the sector and/or the more substantial the externalities it generates, the more likely it is that state ownership of capital will result in a state-socialist firm instead of a public corporation. The reason is simple: if a given type of operation cannot be profitable, autonomous public companies cannot exist. Losses would either take them out of business,[4] or, if the state covers the losses permanently, the companies are unlikely to remain autonomous and must sooner or later turn into state-socialist firms. It is difficult to imagine that the state would continue to underwrite losses without taking part in decisions about the firm's operations. Existence of nontransitory losses would thus entail at least the drafting of plans—approved by the state—outlining the acceptable level of losses in the future, and targets with respect to output, employment, and investments. The firm becomes part of the state sector, not much different from its counterpart in state-socialist countries (though possibly with fewer targets to obey).[5] We would also tend to find the state sector in areas where government policy, for whatever reason, is to keep output prices below their market level (the (a) type of state intervention). This type of intervention, however, is more common in underdeveloped economies, principally because of the low level of income of a significant proportion of population.

Self-employment (i.e., cooperative–1 type) is mostly present in agriculture and services. By its very nature this is a mode of production that does not allow for development of large-scale production. Since hired labor does not appear (except in special cases, such as during the harvest in agriculture), owner-worker firms are technologically feasible only when either (a) production can be efficiently carried with an extremely high nonlabor/labor ratio, or (b) individual entrepreneurial or other special abilities are so important that the production requires very little capital. Situations (a) and (b) are obviously very different. We thus find owner-worker firms either in very capital-intensive areas of production such as agriculture in developed capitalist countries, or in (physical) capital nonintensive activities such as the restaurant business, repair services, consulting, legal and medical partnerships, etc. In the latter, it is the specific qualities of the owner(s) that are essential. Recent technological progress has helped self-employment. While only a decade or two ago its importance seemed to be on the wane, recent developments that have emphasized individual skills, the ability to cater to highly specific demands of a market segment, and flexibility have expanded the area of comparative advantage for the self-employed. Yet, because of its very nature, which excludes (permanently) hired labor, the numerical importance of the self-employed labor force will remain limited. In addition, owner-worker firms are best suited for (1) new (experimental) or (2) only moderately successful firms. In the first case, this is because it is relatively difficult to procure external capital for new or experimental undertakings. If the individual entrepreneur's appreciation of the risk is significantly lower (and/or his estimation of the profit significantly higher) than the average, he may not be able to find a financier. The only solution left would be to start one's own business. When the firm becomes more established it may attract external financing more easily and may need to expand by hiring labor. In either case it ceases to be a pure owner-worker firm. The situation when the firm is highly profitable is similar. Its capitalists-*cum*-workers would naturally tend to enlarge the scale of operations and this would necessitate transformation into a capitalist firm. For example, a successful restaurateur will tend to employ external workers (as opposed to his family) to open up a new branch. Examples of such evolution (e.g., Apple, Microsoft) abound, although the saying "many are called, but few are chosen" applies here particularly well. For this reason owner-worker firms are bound to remain important only in activities that are moderately successful.

The main usefulness of owner-worker firms from the social point of view is that they represent a springboard as well as the testing ground for new ideas. The costs of failure are, from the social point of view, limited since little capital and labor is committed, while the pay-off, if the idea proves successful, may be extremely high both socially and for the individual concerned.

The entrepreneurial type of organization shares some features with owner-worker firms. Entrepreneurial firms are mostly venture capital firms. They do employ hired labor (which makes them bigger than cooperatives), but the need to procure external capital limits their practicability. As mentioned above, this is because the risk assessment of a project by an entrepreneur (who, by definition, provides none of his own capital, but solely manages the project) will often be more optimistic than the assessment of a capitalist who is asked to provide the money. The entrepreneurial type of organization thus requires an unusual combination of (a) important market opportunity not perceived by others, and (b) bullish assessment of that opportunity by financiers who do not themselves wish to undertake the project. Furthermore, the entrepreneur's own assets must be fairly minimal, for otherwise he would probably prefer to start his own capitalist firm or be self-employed. More likely candidates for entrepreneurial firms are individuals with nonreplicable and recognized skills (so that they can attract capital) but also with little money of their own. In a pure entrepreneurial firm, entrepreneurial income is received *in toto* by the project manager (entrepreneur). The capitalist receives his fixed return on the financial capital provided. In real life, however, we shall find, particularly in venture capital firms, sharing of entrepreneurial income between the entrepreneur and the supplier of capital.

The only types of firms that do not, to any significant extent, exist in developed capitalism are firms where "pure" labor (i.e., independently of its ownership of capital) exerts an entrepreneurial role. These are cooperative–2 and labor-managed firms.

6.2 The Social Composition

Appropriation of income in the five types of organizations existing in developed capitalism is shown in Table 6.1

From Table 6.1 we can identify the following social groups: (1) workers in capitalist and entrepreneurial firms, (2) workers in the public sector, (3) workers in the state sector, (4) owner-workers (self-employed and in partnerships), (5) capitalists in limited-liability companies, (6) capitalists in joint-stock companies, (7) capitalists in entrepreneurial firms, (8) managers in joint-stock firms, (9) managers in the public sector, (10) entrepreneurs, and (11) the state.

We must first explain the rationale behind the composition of some of the groups. Workers belong to three social groups, depending on whether they are employed by the private, public, or state sector.

Table 6.1

Appropriation of Income in Developed Capitalism

Type of firm	Type of income			
	Labor	Capital	Entrepreneurial	Taxes
Capitalist				
Joint-stock	Workers	Capitalists	Capitalists/Managers	State
Limited-liability	Workers	Capitalists	Capitalists	State
Public corporation	Workers	State	Managers	State
State-socialist	Workers	State	State	State
Owner-worker	Workers	Workers	Workers	State
Entrepreneurial	Workers	Capitalists	Entrepreneur	State

Capitalists are classified into three groups on the basis of their level of involvement in decision-making. First, capitalists in limited-liability companies who directly manage the process of production. Second, capitalists (i.e., shareholders) in joint-stock firms whose managerial day-to-day role is delegated to managers. As explained before, we assume that capitalists in joint-stock companies retain an indirect managerial role, reflected in the fact that they have the power to decide on the overall management strategy either by voting the management team in and out, or by selling and buying shares. Our analysis in essence assumes that no general strategy of the firm can be implemented without direct or tacit support of the shareholders. Capitalists in joint-stock firms thus as yet do not appear as rentiers, which they do in entrepreneurial firms (our third group of capitalists) where all the managerial functions are relinquished to entrepreneurs. The classification of other groups (owner-workers, entrepreneurs, and two types of managers—in the private and public sectors) is self-evident.

There are three social groups whose incomes are derived from state revenues: the political bureaucracy, state administration employees, and transfer-income recipients. The origin of the last two groups is the same as in state socialism. Political bureaucracy here represents a single social group.[6] A different system of elite-formation (compared to state socialism) makes the distinction between high- and middle-level bureaucracy irrelevant. In effect, we assume that political bureaucracy is formed on the basis of elections in which the whole population participates (or, at least, may participate). All social groups theoretically represent the constituency of political bureaucracy, unlike in state socialism where this role is reserved for the middle-level bureaucracy. This also means that no separate social group exists on whose support top political bodies depend. Politicians will, of course, depend on the support of different lobbies. In a Schumpeterian description of the democratic process, these lobbies will vie with each other to impose their candidates. Yet since the whole population participates in the electoral process, the choice of political leaders will not be determined by competition within *one* social group (*viz.* middle-level bureaucracy) but by competition among *all* the groups. To make clear what we have in mind here, let us note that the situation in socialism is similar to that in feudalism where the power of the king (and often his election) rested solely with the nobility and clergy. Other social groups were excluded from direct participation in the political process. The introduction of formal voting rights and the extension of the franchise gradually made all the citizens part of the political process.

Figure 6.1 shows the distribution of income among different social groups.

Excursus: Political Elites and Freedom of Action

The difference in the ways political elites are formed and rule in socialism and capitalism explains why, in democratic capitalism, the political elite may sometimes have a greater scope for freedom of decisions, for example if the same party controls the executive and legislative branch. In socialism, the same party obviously controls both. But since only one party exists, it is unavoidable that more heterogeneous interests will be present within it than within a given party in a free multi-party system. In state socialism the decision-making within the party thus becomes more consensus-based (there are more constituencies to be satisfied) than in a democratic society where a single party holds the clear majority. The latter disposes of a mandate to execute an equally clear political line and to satisfy a relatively well-defined constituency.

In other words, it matters little to the British Conservative government if say, the blacks in Britain are unhappy with it. As long as they are not strong enough to create social instability and the ruling Conservatives have a strong majority, their views are, for the current policy-making purposes, irrelevant. They do not represent the natural constituency of the Conservative Party. The party can win without them, and has practically written them off. Having no voice, they do not constrain the government's freedom of action. In a single party system, they would almost certainly have some representation in decision-making bodies. Although, due to the political dictatorship, they will not be allowed to air their disagreements openly, their views will be taken into account in the actual policy design.[7]

102

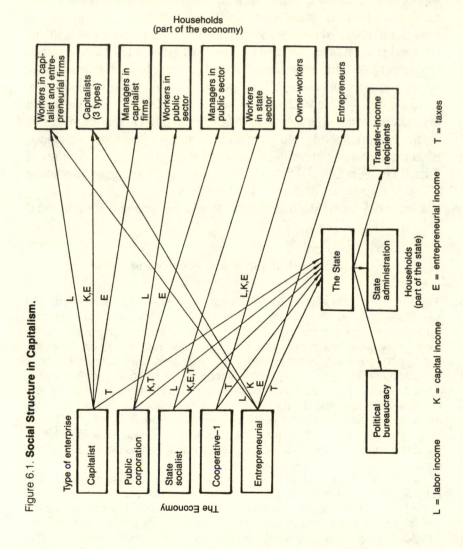

Figure 6.1. **Social Structure in Capitalism.**

L = labor income   K = capital income   E = entrepreneurial income   T = taxes

If we compare the social structure of state socialism and developed capitalism we can already make some preliminary observations. In both socialism and capitalism there are three groups of workers. Two of them are the same: workers in capitalist and state firms. The third is workers in the labor-managed and the public sector, respectively, in socialism and capitalism. Income of all workers except those in the state sector is *grosso modo* determined by market forces. Non-market determination of wages for state-sector (but not necessarily for public-sector) workers also holds in capitalism. The public sector displays a decentralized coordination and operates in an environment where market forces (*via* the influence of the capitalist sector) are dominant. Its rules of behavior must also approximate those of the capitalist firms. The rules of the state sector, however, in matters like wage formation, job security, pensions and medical care, vacation, work discipline, etc., are generally different from those of firms operating under market conditions. [8]

There are three groups of capitalists in capitalism versus only one in state socialism. The capitalist structure is, of course, more diversified: absent in state socialism are shareholders and pure "rentiers"—those who provide capital to entrepreneurs.

Particularly interesting and revealing of the nature of the systems is the difference in social groups that derive their income from the state. While in capitalism incomes of managers (in both private and public firms) are market-determined, in state socialism managers' incomes come from state revenues. [9] This is because managers operate within a hierarchical, centralized system. As Table 6.2 illustrates, five social groups in socialism, but only three in capitalism, derive their (main) income from the state. Conversely, nine social groups in capitalism versus five in socialism have their income broadly determined by market forces.

6.3 What Is Liberalization?

In chapter 2 we saw that liberalization in capitalism is possible within the framework of the capitalist system. Liberalization concerns three aspects, corresponding to the three types of state intervention in the economy. The first is the regulatory aspect. Economic freedom of individual agents expands, for example, by the lifting of price controls, financial deregulation, removal of production licenses, elimination of minimum wage and other labor legislation, etc. This expansion most

Table 6.2

Origin of Income of Different Social Groups

	Market	Non-market	State
		Origin of income	
Socialism	Self-employed Workers in capitalist sector Workers in labor-managed firms Capitalists	Workers in state sector	High bureaucracy Middle bureaucracy Managers State administration Transfer-income recipients
Capitalism	Workers in capitalist sector* Workers in public sector Capitalists (all types) Entrepreneurs Self-employed Managers†	Workers in state sector	Political bureaucracy State administration Transfer-income recipients

*Also includes workers in entrepreneurial firms.
†Two types of managers: private and public sector.

directly affects privately owned and public-sector firms. The situation is different with the state sector. It is quite conceivable that greater freedom from regulations is not accompanied by a reduction of direct state management of the state sector, nor by a shrinking of that sector.

The second aspect relates to the policy on subsidies and protection. Reduction of private sector bail-outs, elimination of subsidized credits, lowering of tariff rates on imports, etc., will toughen the business climate and make economic agents more responsible for their decisions. As for reduction of subsidies to the state sector, we must distinguish between *ex-ante* and *ex-post* reduction. If the state decides *ex ante* on a reduced amount of subsidies for individual state enterprises (and adheres to that), there is clearly a hardening of the budget constraint, less direct state control (which is often correlated with the size of subsidies), and possibly the beginning of a formation of horizontal inter-enterprise links. The latter implies reduction of the state sector and thus liberalization. However, if reduction of subsidies is an *ex-post* reduction, that is, the product of improved efficiency of the state sector, the fundamentals of state involvement in the economic arena remain unchanged. In this case one cannot speak of liberalization; there is only a more efficient state sector.

In effect, liberalization of the state sector takes place either by reduction of state involvement there, or by the shrinking of that sector itself (e.g., simply by a sale of assets). The two are often related. As state involvement diminishes, and horizontal links between the state firms and other firms belonging to the public or capitalist sector emerge, the essential features of the state sector are gradually lost. In general terms, therefore, liberalization of the state sector is inseparable from its reduction.

The third aspect of liberalization is straightforward: reduction of taxes.

In conclusion, *liberalization* in capitalism involves the following actions:

(1) Less regulations for the private and public sectors.

(2) Less subsidization and protection of the private and public sectors.

(3) Reduction of the state sector.

(4) Reduction of the tax burden.

Reform, however, may lead to liberalization in one area (e.g., privatization of one part of the state sector) *and* increased regulation in another.

We shall consider the problems of liberalization relating to the private and public sectors (points 1 and 2), and the problem relating to the state sector (point 3) separately.

6.4 Deregulation

In the part of the economy where coordination is decentralized, liberalization takes place essentially through relaxation or removal of constraints on operations of the firms. These are constraints that relate to prices of outputs and inputs (e.g., government-fixed output prices, legal minimum wage, regulated exchange rate) or to quantities produced or used (e.g., production licenses, statutory limits on number of hours worked). There are also constraints that have to do with the structure of the market, again both on the output (e.g., regulation of the monopoly power) and on the input side (e.g., compulsory syndication of labor).

To clear up possible misunderstandings, it is important to note that our definition of what constitutes liberalization holds even for those cases when the removed regulation may be considered legitimate from the strictly economic point of view, as for example in the case of natural monopoly. We are not giving any value-laden (normative) meaning to the word liberalization. We are using it as a purely technical term to denote situations when state interference in the economy diminishes. Thus any limits imposed on autonomous forms of organization of economic agents, even if these forms are in restraint of competition (as is the case with market collusions or voluntary organization of workers in trade unions), represent state intervention. We consider any spontaneous type of organization of capitalists, workers, entrepreneurs, or consumers to be characteristic of liberal economic order. Put differently: *liberal economic order is not characterized by a perfectly competitive market of atomistic producers and consumers, but by a free organization of economic agents within a given legal framework* (i.e., within defined property rights over factors of production). The concepts of liberalization, competition, and economic efficiency—although related—are not synonymous. They are conceptually different and must, for logical tidiness, be kept in separate compartments.

This leads us to the clarification of a point that concerns the regulation of monopoly. Regulation of monopoly, being an interference with the spontaneous order of the market, represents state interference. On the other hand, type (b) of state interference also included the granting of protection from internal or external competition. Firms that enjoy it represent state-protected monopolies. Consequently, these two different types of monopolies must be distinguished, with the result that state interference in economic life *increases* both (1) when it exercises a regulatory function by *limiting market-created* monopoly, and (2) when it *confers state* monopoly.

Decisions to proceed with the removal of some regulations will be dictated by economic considerations (e.g., if price controls lead to shortages) and political relations among different social groups or, more exactly, between those who stand to gain by the removal of restrictions and those who stand to lose. Removal of restrictions will affect various sections within the same social group differently. For example, nonintervention of the government in the determination of the exchange rate will, if the free exchange rate devalues comparatively to the existing level, be supported by the export-oriented and import-substitution part of the economy and consequently by capitalists and workers in these branches, and will be opposed by importers. Decisions to free some other prices, or to deregulate supply or demand of some goods, will similarly represent redistribution from one set of industries to another. Within the same social class (e.g., capitalists or workers) liberalization will make some better off and some worse off. The implication is that with respect to some types of liberalization the attitude of social groups will not be that of homogeneous blocs. Links of interests will cut across social groups, and we will find workers and capitalists from one set of industries arguing for liberalization against workers and capitalists from another set of industries. The central social conflict in capitalism, which, as we shall show *infra*, concerns the distribution of the net product and opposes workers and capitalists, is significantly attenuated here. The degree of group cohesiveness on questions concerning the regulatory framework in capitalism is much smaller than the degree of group cohesiveness regarding the reform in state socialism. The absence of clearly opposed interests of different social groups enhances the stability of the system.

The identity of interests that exists in capitalism between workers and capitalists in one group of industries against other groups (and the absence of that identity between workers and technocrats in a given industry in state socialism) is the consequence of the existence of the market. In state socialism, workers are insulated from all efficiency considerations. A firm or an industry may be either very efficient or inefficient, but the position of the worker will vary little: neither his salary nor his employment depend on the economic success of the firm, not even on the degree of fulfillment of the plan targets, since there the responsibility is clearly the managers'. As a rule, workers will have little interest in what happens to their firm or their industry. In capitalism, on the contrary, workers' employment and pay are intimately related to the success of the firm. Although capitalists may be the first to gain from excess profits or to suffer from sudden losses, financial improvement or deterioration will quickly trickle down to the workers' level. Their interests thus clearly coincide with those of capitalists and managers in the private sector of the economy *so far as the relative position of the firm or industry (vis-à-vis other firms or industries) is at stake*.[10] Since removal of regulations affects the position of the firm and the industry, there must be a concern among workers for what liberalization implies for their firm, and accordingly a coincidence of interests between them and capitalists and managers from the same firm.[11]

Absence of group cohesiveness on the issue of easing of regulations has another consequence: liberalization will be more easily implemented or reversed, and it will be spotty. Absence of bloc-like actions of different social groups means that, while there may be a sufficiently strong coalition to support measure A, so that it will be implemented, for liberalization measure B there could be an entirely different "winning" coalition that may want to reject it. Some of the "liberalizers" in terms of A may be among the "regulators" when it comes to B. The removal of state regulations in one area may thus be accompanied by the imposition of new restrictions in another area, and the overall drift toward liberalization (or away from it) can be assessed only as the net result of these different moves. There would be a continuous waxing and waning of liberalization. The spotty character of liberalization stems from the existence of fluid coalitions. In effect, if coalitions were stable (even if they cut across different social groups), the domination of one coalition would ensure that most of the measures it favors are implemented. There would be massive liberalization or de-liberalization movements, punctuating the long periods of *status quo*. The reality, however, is different: changes are spotty and piecemeal.

We have thus identified two reasons why reforms in developed capitalism will be both more frequent and less massive than in state socialism: first, because they can take place within an unchanged capitalist framework, while in state socialism they immediately stumble against the dominant centralized mode of production; and second, because coalitions in capitalism are more fluid.

6.5 Reduction of Subsidies and Lowering of Protection

The next liberalization issue is the one concerned with the elimination or reduction of state subsidies to the decentralized (private and public) sector and the reduction of internal and external protection. The situation there is similar to the one that exists with respect to deregulation. Withdrawal of direct and indirect subsidization will be opposed by those who are its recipients, and favored by others who might expect that their tax burden will be lessened or that elimination of some firms would reduce competition. The issue of subsidization may have to be considered in conjunction with other forms of liberalization, namely, the removal of restrictions. If the decision to eliminate subsidization to a firm or an industry is combined with lifting price controls, firms that are to lose subsidies may still support such "linked" liberalization.

The relationship between liberalization and reduction of protection from external markets is not as simple as it might at first appear. Most people would argue that liberalization in this area must include lowering of tariff rates and relaxation of nontariff restrictions on imports (viz., a system of quotas and licenses). It might be thought, however, that looked at from the perspective of effective protection, these measures are not necessarily compatible with liberalization. To explain: An industry that enjoys high protection of its output and low protection of its inputs may have an extremely high effective protection (protection of its own value-added). Obviously, producers who supply inputs to that industry (which must compete with low-tariff imports) are the losers. It is therefore often argued that equality in treatment dictates uniformization of effective protection, achieved by equalization of all tariff rates. This line of reasoning at times seems to suggest that the problem is an excessively low tariff on inputs, and consequently differential treatment of different producers. It might then appear—following the same argument—that the reduction of some tariff rates, if it increases dispersion of the effective protection, should not be treated as liberalization since it magnifies the relative advantage enjoyed by some industries. Now, as explained in section 2.1, liberalization, or its reverse, has nothing to do with differential treatment of firms or industries as such. Consequently, we can define all measures that *reduce* tariff rates as liberalization, even if they lead to an increased dispersion of the effective protection.[12]

6.6 Scaling Down the State Sector

Issues to be considered in the liberalization of the state sector have much in common with those considered in chapter 5. In capitalism, however, devolution of the state sector can take the form either of privatization or of a move toward the public-corporation type of enterprise. In the past, attempts to remedy inefficiency of the state sector, while preserving state ownership of capital, involved the break-up of the state sector into a number of autonomous public corporations. More recently, a more radical approach, namely taking over of state (or public) enterprises by the private sector, is becoming widely used. The state is thus forgoing not only its entrepreneurial role, but also the ownership of productive assets. Motivation is threefold: (a) to provide extra sources of revenue to the treasury, (b) to improve efficiency of the firms by handing them over to the private sector, and (c) ideological preference for the private sector. It is too early to say how successful the process will be in improving the efficiency of the erstwhile state firms, or, in some cases, how successful the state will be in finding buyers at reasonable prices. Yet it is undeniable that the process is becoming global, covering not only developed but also less developed capitalist economies. In terms of our analysis, sale of state firms certainly represents liberalization; sale of public corporations is more ambiguous because it is not always evident that predominantly private ownership of capital will imply less state interference. It is, in effect, quite conceivable that privatization may be accompanied by more extensive regulations or increased protection from competition. The impact of privatizations on the actual level of state intervention in the economy therefore needs to be assessed in each individual case.

A related problem concerns the relationship between privatization and competition. This is of particular relevance in countries with "thin" stock markets where large-scale privatizations, owing to the dearth of small investors, usually mean that firms end up being bought by a few powerful industrialists. This, instead of increasing competition, in fact reduces it. The objective of efficiency improvement, which underlies privatizations, is generally associated with increased competition. To the extent that less, not more, competition is achieved, it jeopardizes the rationale of the whole operation. The same is true if the government, in an effort to maximize revenues, sells a monopoly instead of breaking it up.* An unchecked (or poorly checked) monopoly would, of course, be worth a lot more than a firm subject to the rigors of competition. But merely turning state monopolies into private ones (as was done with British Telecom and British Gas, the two largest privatizations in the UK, and Nippon Telegraph and Telephone in Japan) does not necessarily improve efficiency. Brittain (1986, p. 38), however, has argued that even if the company originally retains monop-

---

*It could be argued that if the company is equally efficient when it is a private as when it is a public monopoly, then the maximum price offered will be equal to the discounted sum of expected net profits, and the overall long-run revenues of the state will not change. The sale will, however, result in increased revenues if (a) the state's discount rate is higher than that of private investors, and/or (b) the firm becomes more efficient under private management, either by better capturing the monopoly profits or by lowering its operational costs. In the second case, one portion of the expected increased returns may be reflected in the private investors' offer price and raise it above its previous level (at which the state is indifferent between keeping the firm and selling it).

oly power, political pressures on the government to allow free entry will increase after privatization. He ascribes this to the fact that firms assume that the government will not protect a new private monopoly the way it did the old state-owned one. Differently, there may be some instances when privatization, in order to be successful, requires less competition. Cranston (1987, p. 278) mentions the case of the public-ly-owned Australian National Line, whose activities were curbed in the 1950s as the government tried to open the area for private shipping.

A special kind of privatization takes place when shares are sold to the firm's employees. In instances where the initial share price is underestimated this represents an important windfall gain for buyers.[13] In terms of our classification, it also means that if all shares are held by workers (and all workers have some shares), the state-sector firm is transformed into cooperative–1. The best known example of employee buy-outs is the National Freight Corporation in Great Britain. When only one part of shares is worker-owned, the new firm lies somewhere between cooperative–1 and a joint-stock company.[14] To the extent that workers gradually sell shares to external investors, the enterprise will evolve in the direction of a usual joint-stock company. Recent experience indicates that such sale of shares seems likely, in addition to "externalization" of some shares by the natural process of attrition of the work force. The scheme therefore does not fundamentally differ from the usual ("outside") privatization. It does, however, represent a wealth redistribution in favor of employees, and they will have an interest in arguing for it. It does not, however, imply a transformation of the state sector into cooperatives; unlike in cooperatives, an employee can easily sell his share and remain in the firm merely as a worker.

Employee ownership (of at least part of a privatized company) also responds to the political interest of the government to increase the number of private shareholders. The greater the number of shareholders, the greater the resistance to possible renationalization will be, and the less likely people will be to vote for parties that promise to undertake it.[15] The policy becomes less reversible, and electoral chances of the party that carried it through—even leaving aside ideological preferences for the private sector that may develop among new shareholders—improve. One of the ways to achieve wider ownership consists in offering shares for sale to a large group of qualified buyers at less than the market price. Rationing then allows each individual to buy only a limited number of shares. Table 6.3 shows that for selected British companies that were privatized, the immediate gain realized by suc-

Table 6.3

Offer and Opening Prices

	Offer price	Opening price	Gain (%)
Amersham	142	188	32.4
Assoc. of British Ports	112	138	23.2
British Aerospace (1981)	150	171	14.0
British Aerospace (1985)	375	420	12.0
British Airports	245	291	18.8
British Airways	125	169	35.2
British Gas	135	147.5	9.3
British Petroleum	363	367	1.1
British Telecom	130	173	33.1
Britoil	185	207	11.9
Cable and Wireless (1981)	168	197	17.3
Cable and Wireless (1985)	587	590	0.5
Jaguar	165	179	8.5
Rolls Royce	170	232	36.5
Trustee Savings Bank	100	135.5	35.5
Weighted average			*21.1*

Source: Calculated from Vickers and Yarrow (1988, p. 174).
Note: Offer and opening price in pence.

cessful applicants, measured by the difference between the offer price and the value of shares at the opening (first) day of trading, on average amounted to 21 percent of the shares' offer price. The sale of assets to its users at less-than-market price has similar effects. The best examples of such "widening" of shareholders are sales of council flats and Telecom shares in the United Kingdom.

The combination of cut-price privatizations to the public at large, employees, or users of particular services, contributed to a dramatic rise in the number of individual shareholders. Between 1979 and 1987 the total number of shareholders in Great Britain expanded roughly fourfold: from a little over 2 million to 9 million. The latter figure represents 21 percent of the adult population, only 6 points below the level in the United States (a paramount shareholders' country). Even if employee ownership relatively quickly becomes transformed into usual joint-stock companies, the number of small shareholders is likely to stay permanently higher than before privatizations: it was estimated that of 7.8 million small shareholders who bought shares in British

Telecom, British Gas, and British Airways, 5 million still held shares at the end of 1987.[16] Broadly spread privatization may, by exposing a part of the population to the idea of shareholding, stimulate a long-term interest in ownership. It was noted, in effect, that successful early privatizations (and, it is true, a buoyant market) in the UK aroused interest among small—often first-time—shareholders to buy stock in regular (i.e., not newly privatized) companies.

We have noted above the three principal objectives of privatizations: (a) revenues, (b) efficiency, (c) ideology and politics. We can now see how they often conflict. Privatizing a monopoly will maximize state revenues, but will not improve the economic efficiency of the firm— that objective would have been better served by increased competition. If the government then attempts to regulate the new private monopoly by establishing regulatory bodies, more competitive behavior may be stimulated, but at the cost of yet another intervention in the economy. Similarly, the political objective of expanded shareholding will conflict with revenue-maximization: the government may have to resort to cut-price privatizations.

6.7 Capitalists and Privatization

The interests of different social groups with respect to the dismantlement of the state sector differ sharply. Capital-owning social groups will generally favor privatization because it naturally expands the private sector and gives them additional areas for investment, control, and profit-making. They will particularly support privatization of state enterprises (or parts thereof) that are likely to be profitable. As for the other enterprises, the capitalists' position is ambivalent: on the one hand, for reasons of principle, they may be in favor of a wholesale privatization, even if individually they are not interested in investing in firms that are sure loss-makers; on the other hand, they may wish that the state keep these firms and absorb the losses, particularly when the enterprises generate substantial externalities.

This second aspect probably illustrates the key ambivalence in the capitalists' attitude toward the state. Ideologically speaking, capitalists must favor a small, so-called "night-watchman" government—ideally, a government whose role is limited to protecting the property rights structure that is most congenial to capitalists' interests (i.e., legal freedom of labor and private property of capital). For that protection capitalists are willing to share one part of their income with the state, viz. to pay taxes. Once the state's role expands, as it did in the last fifty years, to cover direct involvement in the economy and sometimes outright protection of the workers' interests,[17] in addition to protecting property rights, the capitalists' attitude concerning these changes must, strictly speaking, be negative. However, there are economic areas that generate important external benefits though they are hardly profitable in themselves. It is in the capitalists' interest that the state intervene there. The basic negative ideological position toward state intervention in the economy is thus modified. The optimal state policy, from the capitalists' point of view, becomes: minimal state *plus* some involvement in directly nonprofitable areas with substantial externalities.

Now, it should be clear that this represents a direct claim that the state should protect the capitalists' interests. On a theoretical plane, it is probably a departure from the central bourgeois tenet of political and economic equality. This tenet requires that the state does not explicitly encourage or penalize any social group or individual. It should simply protect the existing property structure. The tenet is thus fully consistent with the idea of the minimal state. But with minimal state *plus* some *specified* state involvement, we are really saying that the state should cease to be neutral, and do whatever is in the interest of the capitalist class.[18] This is pregnant with a number of implications: the state would then really become an organ of the capitalist class, as some Marxists argue, and the democratic ideal of equality would be severely damaged. This is why the capitalists' argument for an extension of the minimal state in some specified areas is intellectually unsatisfactory.

A way to get out of this position lies in a redefinition of property rights. In effect, if the existence of externalities simply denotes the fact that property rights are not sufficiently widely spread, then, under a more adequate definition of property rights, most externalities would disappear and the capitalists could reiterate the classical claim for a minimal state alone, without any adverse effect on their economic interests.[19]

If the government decides to proceed with privatization of the state sector, this may give rise to another phenomenon of the symbiotic relationship between the state and capitalists. This is the problem of phony privatizations. They occur when state enterprises are sold far below their market price. Investors reap windfall gains. While this does represent a transfer of assets from the state to the private sector, it is rendered possible only through special privileges granted by the state to investors who are allowed to buy the firm. Levers of state power are used for private gain. Strictly speaking, this is no different from privatization in which shares in firms or state-owned apartments are sold at below-market prices. The impact of windfall gains there, however, is attenuated by two factors: first, the beneficiaries are generally low-to-middle-income groups, and second, the average gain per household (or per individual) is modest because each household (or worker) is allowed to buy only a few shares.[20] Neither of these two things is true for privatizations at below-market value in which big investors receive windfall gains. The problem is exacerbated in a "pathological" case of a cycle of privatization, nationalization, and reprivatization (as in the Philippines) where rich families, using political connections, have been able to buy state enterprises at low cost, then (when they have become unprofitable, often as a result of the owners' mismanagement) to sell them back to the state at unrealistically high prices, and eventually to buy them again cheaply.

In terms of liberalization of the economy, privatization of assets at below-market price has an ambiguous effect: on the one hand, it represents liberalization, as does any movement from the state to the private sector; on the other hand, such massive windfall gains are equivalent to state subsidization of private individuals. We defined the latter as being one of the forms of state intervention in the economy.[21] Our analysis shows how the exact terms under which privatization of the state sector takes place are extremely important; not only because of the injustice that the transfer of assets at below-market value implies for those who are not allowed to bid, but also because it need not really represent liberalization even if it masquerades as such. This can have detrimental political effects when the state eventually decides to proceed with genuine liberalization of the economy. It is unfortunate that phony privatizations are sometimes hailed as indicators of a more liberal economic policy.

6.8 The Position of State-Sector Managers and Workers

Managers of firms belonging to the state sector will, in all likelihood, prefer their devolution into the public sector. Managers would thus maintain their current position (whereas with privatization reappointment of the existing management team is less likely) and become independent from the state. They may not be averse to privatization, provided it does not involve change in the management team or the organizational structure. As Kay and Thomson (1986, pp. 18–32) have argued, managers are particularly concerned with avoiding a movement toward a more competitive environment. If monopolistic state or public companies are privatized, they will try to preserve the same monopolistic position for the new firm. Managers will thus oppose a company split-up or the granting of extensive powers to government regulatory agencies. The main attraction of privatization for them resides in greater freedom from legal restrictions on employment and investment policies of the firm.

Finally, for workers in the state sector, both privatization and the public corporation are less preferable than the *status quo*. For the reasons explained in chapter 5 (more exacting job discipline, redundancies, loss of privileged social security and health policy, greater wage differentiation, etc.), they would be opposed to the change. It is true that in the longer run some particularly good workers may benefit from a more competitive environment. However, a person's attitude toward the desirability of a given course of action is essentially determined by his perceived short-term gains and losses. In addition, state firms in capitalism operate in a competitive environment that is already quite likely to have sucked in the best workers, to whom the private-sector firms can offer better wages. Workers in the state sector who may profit through exposure to increased competition will accordingly be few.

We may expect that the opposition of workers to the devolution of the state sector will be greater in capitalism than in socialism. This for two reasons. First, as explained above, the state sector in socialism contains a greater percentage of good workers who may welcome increased competition. Second, the alternative of the labor-managed firm, which was the key reason for workers to support reform in socialism, is practically nonexistent here. It is excluded in capitalism because of the natural tendency of firms where labor exerts an entrepreneurial role to evolve in the direction of capitalist firms (so that labor-managed firms are heavily marginalized), and also because of the lack of attractiveness to the present owner—the state—of a transition to the labor-managed type of firm. In effect, for privatization there is both political support (conservative parties will favor it) and financial advantage (the state will receive the proceeds of the sale). For devolution in the direction of the public-sector firm, there is again definite political support (generally from the social-democratic parties), and the model of the public-sector firm is the one with which the state is familiar. In contrast, for the labor-managed firm, there is no political support (in the West, no

major party or trade union favors it), and the very type of organization—partly because of its marginality—is associated with charitable or
nonprofit activities. This is certainly the last thing into which the state,
in search of a more efficient model of organization, wants to transform
the firms it owns. In addition, a number of problems may be expected
to plague such labor-managed firms (e.g., lack of access to private
credit) which, in order to be corrected, would again necessitate state
intervention, thus undercutting the very objective of the reform. Finally, if the viability of labor-managed firms in a capitalist environment is
limited and they tend to evolve into capitalist firms, the obvious argument is then: since the labor-managed firms will, if successful, transform themselves into capitalist firms anyway, why not let them become
capitalist right away?

The only thing that can mollify workers' opposition is the promise of
significant participation in the original issue of cut-price shares. The
windfall gain thus realized can be sufficient to compensate for the
anticipated later loss (fewer jobs, stricter work discipline, etc.) that
follows upon privatization. But this presupposes that the government
makes two decisions it may not be willing to make: consciously to sell
the enterprise at less than its market value, and to sell most of it to the
firm's employees.

6.9 What Future for the State Sector?

On the issue of reduction of the state sector we thus find sharply
delineated positions of different social groups. This contrasts with the
situation in other areas of liberalization, where, as we found, support
or opposition to a liberalization measure cut across different social
classes.

Table 6.5 shows the preferences of the three key players in the
decision on devolution of the state sector.

Capitalists and state-sector workers will display exactly opposite
rankings of preferences. Managers in the state sector favor the public
corporation, and may marginally prefer the capitalist firm over the
status quo. If these three groups are regarded merely as three voters
(with equal voting weights), it is clear that the public company solution
would prevail: in effect, it would beat the other two proposals by a 2-to-
1 majority.

The actual form of devolution is, however, more complicated and
will depend on the overall *rapport de force*. Private-sector managers[22]

Table 6.5

Rankings by Different Social Groups of Alternative Organization of the State Sector (preferred ranking = 1)

	Capitalists*	Managers	State-sector workers
Type of Firm			
Capitalist	1	2	3
Public corporation	2	1	2
State-socialist	3	3	1

*Also includes private-sector managers.

will most likely take the same position as capitalists, because expansion of the private sector would enlarge their domain of action. The position of private-sector workers (since they are not directly affected) would probably be neutral. The alignment of social forces will therefore be as follows: privatization will be favored by capitalists and private-sector managers; evolution toward the public corporation, by state- and public-sector managers; and maintenance of the *status quo*, by state-sector workers (unless they are "bribed" by the promise of windfall gains). Other groups may be considered neutral. The position of politicians will depend on who their constituency is. Naturally, conservative politicians will side with the private sector; social democrats, with technocrats in the nonprivate sector, or, if they are more to the left, with workers in the state sector.[23] To the extent that the last group is not influential and the predominant feeling is that something ought to be done with state enterprises, only two camps will eventually emerge: privatization or devolution toward an autonomous public corporation.

Today, when privatization forces seem to be holding sway and privatization embraces not only the state but also the public sector of the economy, it is easy to argue that the decisive victory is definitely won. But one should remember that the same issue has recurred throughout this century and that there have been times when the solution for the state sector was sought in its transformation into the public sector, and even when the cure for the ills of private enterprise was seen to lie in its nationalization. Although we shall argue (in chapter 7) that the present wave of privatization may represent a longer-term phenomenon, not easily reversed in the years to come, the ultimate triumph of the private sector is not a foregone conclusion. We may still, in twenty or thirty years, witness a new round of nationalizations in developed capitalist economies. Nationalizations may then look (as they indeed did in the late 1940s) to be as much a thing of the future as privatizations do now.

6.10 The Central Conflict in Capitalism

The central conflict in capitalism sets capitalists and managers in opposition to workers. The conflict concerns distribution of net product. Two observations need to be made here. First, the statement that the central conflict involves capitalists and workers is contested by writers on the so-called postindustrial society.[24] They consider the view to be a throwback to nineteenth-century Europe, irrelevant for the present-day developed capitalist countries. This tendency is particularly prevalent in the United States, where class contradictions have historically been weaker than in Europe. This is also true for Japan, where capitalist paternalism has blunted the edge of the class conflict. Or both countries are at the forefront of technological progress. Extrapolating from the special circumstances of these two countries, theorists of "harmonies of interests" argue that the coming of the technological society as such is responsible for the quasi-disappearance of the conflict between capital and labor. Second, it is pointed out that the strength of organized labor, i.e., of the key institution through which workers have historically fought for their interests, is decreasing everywhere.

The response to these two statements must start from the observation that it is necessary to distinguish between the conflict that is implied by the set-up of the system itself (i.e., a structural conflict) and the acuteness with which this underlying conflict manifests itself. Increase in real wages, differentiation of wages between various categories of labor, relative decline of traditional (blue-collar) workers, and a remarkable increase of highly skilled technicians, economists, engineers, and others, i.e., of all those who are popularly called "professionals," and finally the increasing share of workers employed in the state apparatus and in services (where class conflict has generally been the weakest), have all contributed to the fact that class conflict, which was quite evident in the early twentieth century, is much less strong now. The technicians, engineers, economists, and lawyers of today, who have taken the place once occupied by workers, cannot be classified as working class, but rather under the vague label of middle class. On the side of the capitalists, on the other hand, the divorce between ownership and management, the dispersion of ownership not only among the

population at large, but also, for any given individual, of his investments in different firms, has severed that important financial and
emotional link between an individual capitalist and the firm in which he
has a stake. Thus both the feeling of being employed (and exploited) by
somebody and that of owning something are diluted. The acuteness of
the conflict between the two social groups is reduced. In that sense the
theories that we mentioned above are right.

Yet if we are concerned with the fundamental set-up of the system,
namely that capitalism is *by definition* a system where the net product is
subject to market-determined distribution between workers and owners
of capital, the underlying conflict of interests is immediately apparent.
The fact that distribution is market-determined does not obviate the
truth that the product is still divided between these two groups, and that
each of them will try to manipulate the market (and politicians) in order
to increase its share.

The existence of hard budget constraint is a necessary condition for
the emergence of the conflict. Value-added produced by an enterprise is
a known, definite, and limited quantity. The simple fact, then, is that
the larger the share of the product belonging to property-owners, the
less belongs to the suppliers of labor. When the budget constraint is not
hard (as in socialism), value-added is an ambiguous quantity. It is in
fact a fairly "expansive" magnitude, capable of all sorts of manipula-
tions. A given physical output is compatible with a whole array of
actual value to be distributed, depending on negotiated prices, subsidies, or taxes. When workers push for higher wages there is no clear
limit to how much of the product they can appropriate, nor is there any
particular reason why managers or the state would refuse to accede to
their demands. By contrast, in capitalism, trade unions in each enterprise know exactly how far they can go: if they ask for too much they
can either be replaced by other, more pliant workers, or owners, making less than sufficient profits, may decide to fold up the firm, with
workers consequently joining the ranks of the unemployed. There is no
reason why in socialism managers would fight workers' demands tooth
and nail—unlike capitalists, they are not defending their own property.
More likely, they will acquiesce to workers' demands and press state
bureaucracy for higher prices and subsidies or lower taxes.* "Elastici-

---

*The existence of the soft budget constraint explains why the central conflict in
socialism does not concern the distribution of the net product between two definite groups of people.

ty" of the value of net product produced by each enterprise is the reason why independent trade unions in socialism, as witnessed in Poland during the "Solidarity" period,[25] are bound to create an inflationary spiral and destabilize the economy much more than in capitalism: lacking the discipline of the market, operating in an environment of arbitrary prices, and, perhaps most importantly, without a clear link between profits of an enterprise and somebody's income, no one will have an incentive to resist demands for wage increases. The ultimate self-budget constraint is the government and the government can go on satisfying everybody's demands so long as it prints more money. The result is inflation and disequilibrium on the consumer market where available money balances of households consistently exceed the supply of goods at state-controlled prices.

Now let us take the extreme case where it may be seemingly easily argued that the fundamental conflict between the two groups in capitalism has disappeared to the point of vanishing, and imagine a system in which skills and property are so evenly distributed that neither pure workers (i.e., people whose income depends entirely on sale of labor) nor pure capitalists exist any longer. Sources of income (capital and labor) are distributed in the same proportion across all income groups: people with a low total income receive, say, 30 percent of total income from property and 70 percent from labor, and the proportions are exactly the same for those with high incomes. (Note that the size of income distribution need not change much.) In other words, the link between the *level* of overall income and the *source* of income (which, particularly in the past, made us associate high incomes with ownership of property) is severed.* Unlike in, say, nineteenth-century Britain, we

---

*Clearly, an empirical analysis of the "classness" of a capitalist society can be conducted using the same model. A society could be said to be classless if relative shares of property and labor income are equalized across income groups or, put differently, if relative shares are independent of the level of income. To the extent that labor (or capital) incomes are located among particular income groups, a society is more class-divided. In one extreme case (classless society) the Gini coefficients for labor and property income, as well as the rankings of all individuals according to their labor and property income, are the same as the Gini coefficient and the ranking of individuals according to total income. In another polar case (sharply class-divided society) the two Gini coefficients for the sources of income are unrelated to the total Gini coefficient (for all income), and the rankings are exactly the reverse: all individuals with a positive labor income will have zero property income, and vice versa.

could no longer *physically* identify capitalists and workers; everyone would become part capitalist and part worker. The class of rentiers, which is always socially destabilizing because its lack of effort stands in such sharp contrast with the income it draws, would no longer exist. Although at present no developed country comes near this model, the spread of highly paid technical jobs (which are, strictly speaking, wages), as well as the dispersion of ownership, seem to point to an evolution in that direction. High wages mean that capital-owning classes will not look at the very idea of working with contempt, but will be, on the contrary, quite keen to snatch up these well-paid jobs, whose wages may exceed their property income severalfold.[26] It is not inconceivable that within the next fifty years mature capitalist societies could resemble that model. Even then, *within* a given firm the conflict between people who receive labor income and those who receive capital income will remain.[27] A person may receive property income in firm A, and there side with other capitalists in favor of increased corporate savings or dividends, while simultaneously, in firm B where he is employed, he may demand higher wages, together with other workers.

The characteristic of capitalism is that it makes the clash between the opposing interests obvious. People who receive dividends and those who receive wages in any given firm are unambiguously two different groups. This is not the case in socialism, where workers do not have in front of them people whose economic interests run as clearly contrary to their own.[28]

As we saw in chapter 5, the central conflict of interests in socialism is not the one between workers and bureaucrats. This is due, first, to the insulation of workers from the market; second, to the fact that bureaucrats receive their income not directly but by intermediation of the state; and third, and most important, to the fact that the key role of bureaucrats in socialism is to check the power of technocracy, and not to make sure that workers are working well and are paid no more than what the market will allow. The *encadrement* role is exercised, if at all, by managers—which by the well-known rule "the enemy of my enemy" makes workers view the bureaucracy rather sympathetically. Capitalists obviously cannot take such a detached view of the efficiency of the firm, because their income directly depends on its performance.*

---

*The "encadrement" role explains why managers in private firms can be included alongside capitalists as a party in the central conflict. First, because of their

We thus see that the very reason that explains the greater efficiency of the capitalist vs. the socialist type of organization (direct linkage between a firm's results and rewards) is also responsible for the existence of that fundamental conflict between workers and capitalists. This is what we meant by saying that the conflict is implied in the system's set-up.*

The intensity of the conflict, and in particular its manifestations, will be different in modern developed societies from what they were in early capitalism. Educated and well-off people will not fight the same way as half-starved miners fought the police cavalry. The reduced importance of organized labor indicates that acuteness of the conflict is less, but also that there may be new ways to fight it. Skilled workers, and in particular "professionals," resent the bureaucratic machinery that the modern trade union has become. They generally decide not to participate, and it is very likely that with the increase in proportion of highly skilled labor the importance of trade unions will continue to decline. But, as indicated above, this affects the modalities of the conflict, not its very existence—lodged in the set-up of the system.

---

position in the process of production, managers must be concerned with the most efficient way of organizing production. This will make them—even if they have no financial stake in the profit of the firm—mentally much closer to capitalists than to employees. This is true even if no class differences are present: in terms of income and social status some employees (say engineers, economists, lawyers) may not stand any lower than managers. Second, the tendency of managers to espouse capitalists' interests is reinforced by their participation in the entrepreneurial surplus, as well as by the fact that shareholders represent the managers' constituency. They must be sensitive to the views of the constituency if they desire to remain in their jobs.

*This shows that the conflict between workers and capitalists cannot be considered merely another instance of the "fundamental" conflict between saving and consumption. The latter conflict is not a "natural" central conflict in all societies, as is sometimes argued, with different social groups in different institutional set-ups taking pro- or anti-savings roles. Capitalists and bureaucrats would, according to this view, take a pro-savings role in, respectively, capitalism and socialism, with workers always being pro-consumption. This view is incorrect for the two following reasons. First, in socialism neither the technocrats nor the bureaucrats between whom the central conflict occurs can, as a rule, be identified with a pro- or anti-savings position. Second, since it is not at all certain that the average propensity to save out of dividends is greater than the average propensity to save out of equally high wages, a conflict regarding the use of discretionary income (whether to distribute it as dividends or wages) cannot be subsumed under the rubric "saving versus consumption."

6.11 Would Elimination of the Central Conflict in Capitalism Lead to the Disappearance of the Capitalist Mode of Production?

Theoretically speaking, the central conflict in capitalism could lose most of its meaning in a society that satisfied the two following conditions: (1) approximate proportionality of labor and capital incomes across income groups, i.e., wide dispersion of property, which, as we said, could be realized in a capitalist society of the future, *plus* (2) a declining share of property income. This second condition would necessitate very high capital saturation, so that the marginal product of capital approaches zero. The labor-capital conflict would be devoid of meaning since it would involve a very small part of the total income. The question, however, is not so much whether such a society could ever be realized (there may be reasons to believe that it could), but rather whether such a society would be compatible with the preservation of the capitalist mode of production as the dominant one. In effect, in conditions of near zero cost of capital, access to capital would be open to everyone. It is scarcity of capital that originally enabled capitalists (as it still does) to take the role of organizers of production. If everyone could draw as much capital as he liked, there would be a natural tendency for individuals or groups of individuals to take over the entrepreneurial role, and the capitalist mode would be displaced by self-employment, partnerships, or even cooperative-2. The very success of capitalism, by creating conditions for the solution of its central social conflict, would pave the way for its disappearance and replacement as the dominant mode of production by other modes where labor would play the entrepreneurial role.

Part III

A POSSIBLE OUTCOME

Chapter 7

The Age of Technocrats?

"*It is quite clear that if the impossible becomes possible, and the improbable probable, if world capitalism led by European capitalism establishes a new dynamic equilibrium . . . and if capitalist production in the forthcoming years and decades begins to expand, we, the socialist state . . . will have to try to catch the express train. To put it simply: . . . it would have meant that we were wrong in our fundamental assessment of history. It would have meant that capitalism has not yet exhausted its historical 'mission' and that the present phase of imperialism does not represent its decline, its last gasp, but the birth of a new era of its progress.*"—Lev Trotsky*

7.1 Introduction

In the preceding chapters we outlined a conceptual framework within which to study the liberalization process, and applied it to the issues of liberalization in socialism and capitalism. This part of the book is of a speculative nature. We shall try to look at what changes in the world order may be expected if the process of reform outlined above takes place in both capitalist and socialist economies. There are two kinds of uncertainties: first, whether the process itself will be successful, and second, if it is, in what way it might affect the world order. Discussion

---

*Lev Trotsky, *Towards Socialism or Capitalism?* London, 1926. Quoted from Gligorov (1984, pp. 76–77).

of these issues, and predictions that one makes, are therefore at best tentative. To appreciate the difficulties with which an assessment of future trends is always riddled, consider, for example, Burnham's *The Managerial Revolution*, published in 1941, in which he predicted—quite reasonably in view of the situation at the time—that the world would be divided into three major powers: Japan, Germany, and the United States.[1] Now, the outcome of the last war seems to have proven him wrong: neither Japan nor Germany (the latter, moreover, divided) is today the political power that it was in the 1940s. Yet, Burnham's predictions are borne out if the economic importance of the three countries is considered. The countries he singled out are indeed the dominant economic powers, at the cutting edge of the economic and technological revolution. In another prediction Burnham (1960, p. 183) suggests: "Does any serious person think that the European Continent is again going to be divided up in a score of sovereign nations, each with its independent border guards, tariffs, export restrictions, currencies, forts, armies, bureaucracies. . . ?" The political fact of European division would seem to contradict Burnham: there are roughly as many states in Europe today as there were before the last world war. Yet, on further reflection, isn't it true not only that "independent border guards, tariffs, export restrictions," but also independent "armies and bureaucracies," have all but disappeared? The first through the economic integration via the European Economic Community, the second through political integration under the auspices of the NATO and the Council of Europe in the West, and the Warsaw Pact in the East. These examples bring out the uneasy and ambivalent relation that a prediction bears to reality.

7.2 The Ascent of the Problem-Solvers

Political power in the major countries of the world is now more than at any point in this century, held by the type of people whom we may call technocrats or "problem-solvers." They symbolize the ascent of *Homo economicus* to the forefront of politics. All social systems increasingly seem to emphasize economic issues and their pragmatic solution. The key international figures of the 1980s exemplify this trend. What Reagan, Thatcher, Gorbachev, Deng, and Rajiv Gandhi have in common is that they are pragmatic leaders who put economic problems at the top of their agenda (to a greater extent than their predecessors) and, in trying to solve them, are led by pragmatism more than by ideology. Obviously, political pragmatism—as our previous analysis tried to demonstrate—does not operate in a vacuum. It must always consider what is politically feasible and must introduce economic changes within a more or less given legal framework,[2] and within a given (inherited) constellation of political forces. Consequently, the exact manifestations of pragmatic economic policies in the Soviet Union or China will differ from those in the United States or Britain. But the five leaders (whose countries comprise more than one half of the world's population) seem to be characterized by a willingness to innovate in economic matters and, what is particularly important for our purpose, to innovate in a way that reduces the role of the state in economic life.

The advent of the (economic) problem-solvers, however, is not an entirely new phenomenon. Indeed, there are arguments that the whole second half of the twentieth century is characterized by the predominance of economics over other areas. The industrialization of the Soviet Union, the development of India, the Chinese "Great Leap Forward," are all—successful or not—testimony to the primordial role assigned to economics. In contrast, when one studies, for example, European diplomatic history in the period between the two world wars, one cannot but be struck by the importance that noneconomic issues had at that time compared with the present. The continuous makings and unmakings of political alliances, the coveting of other countries' territories, political pacts, etc. all contrast sharply with the situation that obtained from the 1950s onwards, when—once the political division of Europe settled—economic issues came to the fore. It is difficult today to imagine a reoccurrence of the Byzantine transactions that went into the establishment of the *Petite Entente*, its position vis-à-vis Hungary, the ambivalent meaning of the numerous treaties and guarantees proffered by the French and English to their East European allies, or the complications of the Genoa, Stresa, or Munich conferences. It seems that—with the exception of the Iberian peninsula—no country's alliances were stable, its frontiers uncontested by neighbors, and the country itself without a claim on somebody else's territory. Shifting alliances, continuous negotiations, military threats, displays of force, all belong to the arsenal of the 1930s, which impresses us by its manifest obsolescence. It is not totalitarianism—now as then very much in evidence—that we may find "*passé*" but a complete absence of the attempt to cloak totalitarianism in economic language. It was the language of soil and blood, nation and state, that the fascist and authoritarian regimes of the age used. Scarcely an attempt, except in Soviet totalitarianism, was made to justify a regime on economic terms.[3] In contrast, it would be difficult today to find any regime, democratic no less than a totalitarian or even a simple personal dictatorship, that does not attempt to project itself as concerned with economic growth, or presiding over the country's economic emancipation, or finally, claiming to be concerned with the equitable distribution of income and the welfare of its citizens. This difference in the type of rationalization offered by totalitarian regimes of the past and the present mirrors the difference in the *esprit de temps*.

Why is the rationale proposed by dictatorial regimes to the public important? Because by studying the justification, the façade that dictatorial regimes present to their own citizens as well as to the world at large, we can best judge the spirit of the times. These regimes—in order to acquire some legitimacy—need more than others to follow public moods, to cater, at least verbally, to what the internationally prevalent perception of the "good" is, and to pretend to espouse the same goals. (Whether they really implement them is another matter.) In stressing the economic impact—be it, as in the less developed countries, where the rule of a domestic elite is justified by evoking the principle of national independence and economic emancipation from the West, or in socialist countries, where the system's claim to legitimacy rests on fast growth and economic equality—the theoreticians of totalitarian regimes reflect the dominant public opinion of the time. If today a regime were suddenly to look for its justification in the realm of the blood-and-soil ideology, it could hardly count on any degree of outside acceptability.[4] Although its citizens may, by sufficient repression, be cowed into obedience, a regime so much out of step with the rest of the world could not expect to survive for long. Even when the existing political regimes are ultra-nationalist, perhaps only a shade less than their predecessors in the 1930s, the nationalism that they present to their own population and to the outside world is mainly an economic one. The hypocrisy of dictatorial regimes thus gives us an easy insight into the ideological fashions among public opinion, an inkling of what is the spirit of the times.

The emergence of economics as the key area around which political battles are waged dates to the period following the last war and the accession to independence of a number of previously colonized countries.[5] Economic discourse initially invaded the domestic politics of these countries, but soon it spread to the world. International forums, politicians, scientists, and journalists alike began to express themselves in the language of economics. Political positions were defended or attacked in economic terms. Thus multinational companies could be decried because they weakened the economies of the host countries, subjected them to unequal trade, or reduced them to the perpetual position of raw-material producers. But they could also be hailed as conduits for the transfer of new technology, management techniques, and economic modernization of the host country. The important point, however, is that the arguments, in contrast to the period before the war, were conducted in economic terms.

What seems new in the recent period, therefore, is not the accent on the economic problem *per se*, but the reaction against the type of economic ideology that held sway throughout the 1950s and the 1960s and still lingered during the (ideologically) "troubled decade" of the 1970s—the ideology of *"dirigisme."* This was the approach that shaped the national policies of many countries and dominated the intellectual arena. In short, *dirigisme* placed the brunt of the responsibility for economic development on the state. It viewed the state as a necessary and essentially benign agent of economic transformation. In the fifties and the sixties *dirigisme* was, as discussed in chapter 4, equally influential in England and France, as in India, and, of course, in the USSR and China.[6] Today, on the contrary, the intellectually prevailing attitude toward *dirigisme* seems to regard the whole approach as having outlived its usefulness.

To say that the *dirigiste* approach has exhausted itself is probably the most accurate description of both its past achievements and its present-day inapplicability. The *dirigiste* strategies (often associated with policies of industrialization at any cost and substitution of imports) were not always inefficient at the time when they were undertaken, nor are the countries that adopted them necessarily worse off today for having done so. But, the objective of this work is not an assessment of the accomplishments of *dirigisme*. What is important for us is that public opinion, which in the past favored strategies that conferred on the state a central role in economic life, has now swung in the opposite direction. If one agrees that the shift has occurred, it is really immaterial for our purpose whether the previous policies (1) were wrong, or (2) were appropriate for that period but are no longer now. What determines economic life and policy *now* is the current belief. The issue of whether the *dirigiste* approach was (1) or (2) is not only without much practical relevance, but, due to the subject matter of economics, is unlikely ever to be solved conclusively. Economists who try to argue (1) are trying to prove an unnecessarily strong proposition: that state-based economic development strategies must be inefficient in all places and at all times. This is obviously a proposition difficult to prove. But, in addition, its proof is not necessary to argue that more liberal economic strategies are preferable *today*.[7] In effect, one can reach an identical policy conclusion by arguing the "weaker" proposition, namely that while *dirigisme* might have been efficient in the past, it has become a hindrance to economic progress in the present age of rapid technological and social change. The simplest argument is that the diffuseness of technical progress, the multiplication of the number of economic agents, and the increasingly dispersed ("grass-roots") nature of economic information make the centralization of decision-making more difficult and more costly in terms of forgone opportunities.[8]

To sum up, it would seem correct to state that (1) recent developments have demonstrated that liberal and decentralized economic policies are more efficient in economic terms than state-oriented policies, and (2) public perception reflects this fact. The second proposition essentially means that the ideological spirit of the times supports a scaling-down of the role of the state in economic life. Its corollary is that the leaders of both democratic and nondemocratic countries will be obliged to reckon with that view. In democratic countries, they will be compelled by the direct action of public opinion, i.e., through elections; in nondemocratic, by a growing realization of the inefficiency of the system and the threat it poses to stability.

The last paragraph opens the whole nexus of relations involving reality, its perception, and the translation of that perception ("popular will") into a political result. The crucial element in this formula is perception. This raises the issue of the "durability" of the spirit of the times. By its very nature, the spirit of the times is fickle. What seems true at one time, may appear to have been an illusion twenty or thirty years hence. When the *dirigiste* dogma ruled, there were few who argued in favor of free markets. It is not simply because intellectual fads change that the liberal view has become dominant: it is because the objective conditions—the nature of technological progress and the limits to growth imposed by the preservation of certain types of property and management relations—have quite obviously made one type of economic policies less efficient than the other. Intellectual developments follow changes in the "hard" facts of economic life, and do not happen without rhyme or reason. One may remember Schumpeter, who in

Capitalism, Socialism and Democracy, published in 1942, viewed monopoly and centralization as dynamically more efficient than pure competition. Similar ideas underlie Galbraith's *The New Industrial State* (1967). There was a time when it was generally thought that bigger was equivalent to better. It is useful to recall these views, the better to measure the distance covered since. It is indeed a long way from holding that big firms and the technostructure are economically more efficient, to the current emphasis on flexible technologies and decentralization in the workplace (e.g., the introduction of group profit centers and abolishment of the assembly line in Volvo factories).[9] This, of course, opens up the possibility that at some future point in time, a reduction of the state role in the economy will cease to be regarded as desirable, and that we will witness either a reversal to state-based economic policies or some new policy mix. Although at the present juncture it may seem that technological progress will—if anything— reinforce the emerging consensus on the need for retrenchment of the state, this must not be taken as a foregone conclusion. No one can predict the nature of technological progress and how it may affect economic relations, let alone economic ideology.

7.3 Unification of Elites

The increasing importance of the problem-solving mentality heralds the rise to power in a number of key countries of social groups that share the same technocratic view of the world. This does not necessarily imply (and often does not imply at all) an agreement about objectives among different countries' elites. The objectives of countries may remain mutually exclusive, and competition among them need not slacken. The area of agreement, however, extends to (1) the way the operation of the world (politics in the widest sense) is perceived, and (2) the emphasis on economic—in contrast to ideological, political, and military—factors. This amounts to a shared *Weltanschauung* that makes communication possible, and to the reintroduction of what may be called a normative unity of the elites.

The world elite has been fragmented since the October Revolution in Russia. It could be argued that before the revolution, the elites of all the major countries (which essentially meant Europe and North America) shared a similar view of the world. This is best evidenced by the personal relations and intermarriages among the members of the elites. It was indeed common in Europe in the second half of the nineteenth century to find people of power and money (or, lacking the two, talent) from all parts of Europe and North America at the same spa or resort or tourist site. The Russians were as frequent visitors to the French Riviera, German spas, and Italian cities as were the English or the Germans. The fact that these elites mingled, talked to each other, danced, and intermarried shows that they shared a similar world outlook and acknowledged each other as *belonging* to the same social group. Cleavages indubitably existed, not only among different nationalities,[10] but also, in terms of social background and the origin of their power, between aristocratic elites from the east of Europe (Germans, particularly Prussians, Hungarians, and Russians) and the more capitalist elites of the west. Yet the merging of the aristocracy with the rising capitalist class, which occurred in every European country that was becoming capitalist, was also happening on a Continent-wide scale.

The making of a unified world elite was not the result of an unusual process, nor was the normative unity of the elites in itself something uncommon. On the contrary, it is the segmentation of the world elite that is extraordinary.[11] One such precedent occurred with the French Revolution, when those who took power as the revolution grew more radical had nothing politically or socially in common with the previous ruling class, nor did they (what is more important) share a world view with any of the ruling elites in Europe. Individual elites therefore no longer opposed each other on the familiar grounds of national power; the French Revolution brought into the conflict a more dangerous and divisive weapon—ideology. It is the universalism of the rules that the revolution upheld, combined with zeal to spread the belief, that threatened the established order. The difference in the language spoken by, say, Saint-Just and Edmund Burke is testimony to that. While the language used by one monarch or another was interchangeable—because it relied on the same set of values—there is no possible translation of the language of *"enragés"* into that of the eighteenth-century defender of absolute monarchy by the grace of God. This fact illustrates a deep split between the elites. Although people who thought like Babeuf, Robespierre, or Saint-Just had existed in France before, and certainly in other European countries as well, they had never, up to that moment, achieved a position of prominence and power in which they themselves became the new elite. Their views thus mattered little. It was this new rule of (what for the old elites was) the rabble—i.e., the very extremism of the French Revolution—which turned a number of initial supporters away from it and alarmed the ruling classes in other countries, much more than the proclamation of the Republic.[12] For although a republic is not a pleasing arrangement to kings and their acolytes, it was not—even at that time—antithetical to the bourgeoisie. Not only did the bourgeoisie play the dominant role in the Constituent Assembly, but there were republican capitalist precedents—in the city states, such as Venice, and also in the Netherlands and the United States. The split between the elites was healed with the accession of Napoleon, who reverted, in the course of time, to the trappings of monarchical power, and whose rule was, in social terms, based on the propertied classes. The war between France and England was not any less real for that, but it represented a return to the traditional pattern of conflicts of national interests. The extreme anguish and fear that accompanied the foreign intervention seeking to overturn the French Revolution were absent. The new war was seen as a clash of interests, not as a Manichean struggle between irreconcilable ideologies with claims to universal validity.[13] And it does not seem likely that the relatively benign punishment reserved by the Allies to Napoleon would have been meted out to someone like Robespierre.

The same split between elites occurred with the Russian Revolution. Power was taken by a group that was extremist in nature and totally outside the normal political spectrum of European (including Russian) politics,[14] and whose ideological beliefs were the exact opposite of the liberal capitalist beliefs held by the ruling classes in the rest of Europe. It was hard to imagine a Bolshevik bathing at Baden Baden alongside a Manchester industrialist and a Prussian Junker. The very unlikeliness of this idea (which shocks us even today) illustrates how the group that took power in Russia was ideologically, aesthetically, and culturally far from the elite in the rest of Europe. It is therefore no surprise that the Bolshevik leaders were haunted by the spectre of "Bonapartism"—a possible reversal of the revolution.[15] Similarly, it is no surprise that *international* intervention, like the one that more than a century earlier hit revolutionary France, now occurred in Russia. The only difference was that the new elite was victorious in Russia, and thus succeeded in effectively splitting the world into two mutually competing camps. The competition was not merely nationalist, but ideological, which only made it fiercer. Communist ideology claimed not only Russia, but the world. Nowhere could the ruling classes feel secure. The Russian Revolution was perceived only as the first installment (and in an unlikely place—which of course only increased the anguish of the elites elsewhere) of the world revolution to come. The break between elites thus took place yet again, and the subsequent victories of the Communists in Eastern Europe, China, and South-East Asia reinforced the cleavage.[16] There were henceforth two different elites, which not only laid competing claims on the world and could not agree on the means by which these claims would be adjudicated, but spoke different ideological languages.[17]

It is this cleavage that the ascent of technocrats may bring to an end. The language used by Gorbachev is not different in substance from the language of an American president. Both prominently focus on improvements in material well-being, on new horizons that their economic policies are to open—in short, on economic progress. Both essentially derive their power (or the consensus that keeps them in power) from success in the economic sphere. Concentrating on top people has only a symbolic significance; what really matters is how deep this process of blending goes. Before 1914, it was not the fact that the Kaiser and the Czar were cousins that was of relevance (although it had symbolic significance). What was important was the intermingling of the commercial and aristocratic classes of Germany and Russia. Similarly, the question now is to assess how deeply the process of "technocratization" may go in Communist countries. For it is principally in these countries that one may expect obstacles to the ascent of technocrats to positions of prominence, and consequently to the unfolding of the process of unification of elites. On the other side, in the West, the process of unification of elites has progressed further than ever in history. It is unnecessary to go down the list of all the interdependencies that link the monetary and fiscal policies, financial and stock markets, exchange rate management, agricultural policies, industrial protection, defense, etc., of the developed capitalist countries. The very fact that we can almost pick at will any aspect of the economy or politics to show that it cannot any longer be defined in purely national terms illustrates the extent of interdependence.[18] For the unification of the world elite it is necessary that the ascent of technocrats continue, in particular in the Communist countries; that the technocrats penetrate deeply within their own societies; that the new elites fundamentally share the same outlook on the world, and that they acknowledge each other as belonging to the same species. In other words, that—unlike in the period from 1917 to the present—they speak the same language. At the entry into the twenty-first century this seems to be a language of technology, computers, economics—in short, a language of *econotechnics*; but what really matters is that it be a shared language, not unlike the one of the bourgeois world at the end of the last century.

7.4 Obstacles to Unification

The ascendance of technocrats and the unification of the elites—if it does take place—would represent unification on "capitalist" rather than on "socialist" terms. Accordingly, the obstacles are more likely to come from the socialist side.

State socialism, in the form in which it exists today, seems to have exhausted its creative potential and to have come to a point where it generally loses in economic competition with a renascent liberal capitalism. In purely Marxist terms, the existing relations of production in socialism stifle further growth of the productive forces. This imposes the need for a substantial change in the economic set-up of state socialism. The necessity to change, of course, does not derive from a pure theoretical realization of the limits inherent in the present organization of these societies, but from the elites' confrontation with reality. In other words, it derives from the inability of the system to generate technological progress internally and to ensure economic growth; from the dissatisfaction of the population, which finds its standard of living stagnant while it rises in capitalist countries; and finally, from the military and political vulnerability—entailed by economic weakness—of the system in its competition with capitalism. The last point, as indicated in chapter 5, introduces the element of external pressure for change, particularly in the case of the Soviet Union and China, due to their status as world powers in direct competition with the United States and other Western nations. Falling behind would have far-reaching effects on the world position of the Soviet Union and China in the next century—even to the extent of transforming them from global to merely regional powers.

Although countries that follow the model set out by Marx may point to some of their social achievements (e.g., generalized health care and social insurance, a relatively low level of income inequality, the absence of unemployment, relaxed work discipline), and argue that the success of socialism should be measured by these achievements and not solely in terms of economic growth, it is the economy that ultimately determines the survival or the extinction of a given social system. It is somewhat ironic that such an elementary Marxian lesson should have been imparted to people who profess to be his disciples. For even if one were to accept that the life-style of the population in a socialist country is, all things taken into account, superior to the kind of life led in a capitalist country, what determines the expansion of one system over another is not the subjective individual preferences of people but economic performance. In other words, even if one might hold that it is better to be a worker in Hungary or Cuba than in South Korea, when confronted with a choice whether to buy products made in one country or the other, he would probably choose Korean products. Following his own individual interest, he is thus contributing to the expansion of the mode of production existing in South Korea (of which he may not approve) and to the shrinking of the one existing in Hungary or Cuba. These are the objective circumstances of economic life that—in a Darwinian fashion—lead to the expansion of the more efficient mode of production.

Less efficient modes can subsist only in two ways: (a) by closing themselves off from the world, i.e., by becoming autarkic, which in turn further increases their backwardness, or (b) by hoping that individual economic agents act against their own interests. Point (b) would translate into individuals buying worse and/or more expensive products simply because they are produced in a country of whose system they approve. Even if such behavior can be found under some unique circumstances (e.g., in the "heroic" period immediately following a social revolution), it is a behavior that is not sustainable in the long run—those who might try to persevere would rapidly run out of cash—and better suits the rules of a religious sect than "the ordinary business of economic life." Consequently, the feasibility of solution (b), which for an economist is a dead end from the outset, is excluded for any larger group of people and for any longer period of time. It is nevertheless significant to point out that—clutching at straws—the opponents of reforms in Communist countries often advocate solution (b) as the remedy for economic problems. For this is precisely what the unwillingness to accept human nature as it is, and the appeal to "consciousness," amounts to. In other words, this is a manifestation of an attitude that Bertolt Brecht noted already in 1953, namely that the system is perfect but that the people are not good enough for it. We thus often find Communist regimes resorting to such gimmicks, harking back to their immediate postrevolutionary past when for some time "consciousness" could rule over economic interest. But exhortations to work against one's own interest cannot be very effective, and such gimmicks are thus undertaken more and more half-heartedly. Consequently, the only remaining solution for the less efficient economic systems is autarky.

Autarky of course means acknowledgment of failure and a resignation from the real world. But since there are social forces that may gain from such an outcome, we shall briefly consider whence the most important obstacles to the opening-up and the emergence of a technocratic elite may come.

The first important obstacle is the nature of the socialist economy, and more precisely of the firms in it. Socialist firms suffer from insensitivity to costs. This is due, as explained earlier, to the existence of the soft budget constraint, i.e., to a justified belief that inefficiencies will ultimately be covered by the state. A system in which rewards for success are slim, and punishments for a failure practically nonexistent, cannot be very efficient economically. But this is a problem of internal efficiency. The problem that concerns us here is that state-socialist firms—because of the nature of property relations, one effect of which is the soft budget constraint—are in a weak position vis-à-vis firms in which capital is privately owned and the owners have fairly good direct or indirect control over the operations of the firm. This is, of course, what we would expect, since profits and losses in capitalist firms directly translate into an increase (or decrease) in the wealth of their owners, who consequently have an incentive to follow the performance of the firms closely. On the contrary, state-socialist firms are normally treated by their employees as not belonging to anyone. The notional owner—the state—is not only remote, but is (as we saw in chapter 3) a nonoperational entity, since it is the incentives for the real people in the center who have some influence on the management of the firm that matter, and these are very slim. In effect, the link between the performance of the firm and the reward of the people who manage it is a very weak one, and is rendered even more tenuous and difficult to implement by the existence of numerous price and other distortions that make the unbiased assessment of performance of the firm well-nigh impossible.[19]

State-socialist firms, due to the nature of property relations and the type of incentives, are likely to be losers whenever they enter into a business contract with privately owned firms. This is a well-known phenomenon in socialist countries where the state-owned firms are either banned or strongly discouraged from having dealings with the private sector. It is important to note that this is a problem separate from the issue of the (often) lower efficiency of the state firms. The latter is dealt with by prohibiting even the small-scale private sector from competing in the lines of production where there are state-owned firms. The problem with which we are concerned here is that the difference in the property arrangements between the two types of firms does not encourage managers and workers of the state firm to protect the firm's interests to the best of their abilities. Since gains or losses have but little effect on their wages, they are generally indifferent to costs, ready to pay suppliers as much as they ask, and in some cases positively eager to obtain the best possible terms for a private firm with which they deal if they have friends or some pecuniary interest in that firm. At any point of interface between the two sectors, there is thus an incentive to transfer state property into private property. Although the same incentives exist in capitalist joint-stock firms, more immediate control by the owners renders such "transfer pricing" difficult. Similarly, more efficient markets make the detection of over- or underpayments easier. In the case of state-socialist firms, control is much more lax, and possibilities for misuse greater. They are expanded by the practice of nominating people with strong political ties to management positions, which further limits their accountability. But the most important fact is that no one has an incentive to look carefully into the accounts of the firm, because no one has anything to gain from it, and possibly has quite a lot to lose (e.g., in terms of harassment by the higher-ups who are interested in covering up a deal, the time spent to prove that an illegal action has occurred, etc.).

This fundamental weakness is common to all state-socialist firms and also to the state sector in the mixed economies. In principle there are two ways of dealing with it: either by discouraging contacts between state firms and the private sector, or by introducing comprehensive state property and a planning system in which sales and purchases of commodities are only accounting transactions and no good leaves the state sector (except in the final stage when it is sold to households). In the latter case the entire economy is practically a single firm, and at no point does contact between the state and the private sector occur.

This fundamental weakness also explains why, in socialist economies, those private firms that are allowed to exist are left free to operate mostly in final goods and services, where they face a private consumer with an equally hard budget constraint, while the "commanding heights" of the economy are occupied by the state sector. This problem objectively limits the spread of the private sector in a number of economic activities, even if the authorities are aware that it could usefully complement the state sector. It also explains why the state sector is naturally "expansionist": not being able to survive in competition or in exchange with the private sector, it can only be saved through expansion, i.e., by elimination of all rivals. We thus see how intrinsic weakness naturally leads to "expansionism," or to what may be called *la fuite en avant*.

Now, while the problem of "transfer pricing" can somehow be "solved" at the national level by limiting the interchange between the two sectors, it cannot be solved at all once international trade with capitalist countries is introduced. Structurally weak state firms face much stronger capitalist firms, and what was before an internal (within the country) "transfer pricing" will in this case acquire the characteristics of a transfer of resources from one (socialist) to another (capitalist) country. This is the first (structural) obstacle to a better economic integration of the socialist countries with the rest of the world.

The problem is further complicated by the absence of legal, marketing, and economic knowledge of the rules that regulate international trade (and which are basically capitalist rules). Firms and people used to dealing in a much more relaxed environment, where lawyers practically do not exist, prices are arrived at through bargaining, contracts only imperfectly observed and rarely enforced, schedules not adhered to, loans seldom repaid, and legal sanctions almost never applied, are bound to be, at least in the beginning, the losers in the new game whose rules they hardly know. They will be bewildered by commitment fees, penalty rates, returns of defective goods, seizure of assets, complicated legal proceedings. But unlike the structural problem of ownership, the knowledge gap is a problem that can be solved after a relatively limited acquaintance with the world (capitalist) market. In effect, the same problem had to be overcome by a number of "newcomer" countries to the world of international commerce—Germany in the nineteenth century, Japan in the early twentieth, and South Korea and Brazil more recently (to name only a few). There is no reason why it should not be solved by state-socialist countries. Yet what remains is that (1) the structural problem of ownership limits the level of interchange that socialist countries can maintain with the capitalist economy, and (2) the knowledge gap will induce additional losses particularly at the time of entry into world trade, which obviously strengthens the position of social groups who oppose it.[20]

The third important obstacle to international integration is the position of the Communist party in a reformed system. This includes, first, the undefined (or undefinable) role that the party would have in a new system, and second, the natural tendency of the system to evolve further in the direction of liberalization (including introduction of capitalist forms of organization) than originally envisaged by the reformers. The last point is of particular importance because it has often provoked the reversal of reforms. In effect, we start from the assumption that reformers' original plans call for transformation of the state sector into a number of autonomous public corporations. But once these firms (and their managers) are given entrepreneurial responsibility and the dependence of managers on political powers-that-be is weakened, there is a natural tendency for the movement to outstrip the original objectives of the reformers. Managers, for example, in an effort to procure investment funds (which no longer come from the center) may consider issuing bonds, shares to workers, or even shares to the outside public (to other firms or individuals). Clearly, if some of these types of financing (e.g., private shareholding) are illegal, they will not be undertaken. Nonetheless, the pressure would exist to move in that direction under one guise or another. Quickly, the managers and politicians will clash again—this clash coming soon after the managers' demands for autonomy have been met and the reform has begun. For the managers will naturally wonder what kind of autonomy it is, in which they have to abide again by ideological rules. It will seem absurd to them that while firms are starved for funds and there are people willing to lend or to invest, the transaction cannot take place. Similarly, they would argue that if they are officially free to make all the decisions, and are alone responsible for the management of the firm, they should be subjected solely to the rules of the market. But, to the bureaucracy, the clash, occurring so soon after it conceded a major part of its power to technocrats, would represent further evidence that reform leads to the erosion of the ideological tenets on which its rule is based. The newly won autonomy of the firms coupled with capitalists' latent reemergence on the scene will threaten to leave the bureaucracy without either a political role or an economic power base. The bureaucracy will realize that what it believed was a concession of one part of its power in order to safeguard the rest turned out in reality to be but a first step on the road to self-demotion. The bureaucracy's reluctance to yield further, nay, its readiness to reverse the concessions it has already granted, will stiffen. This point illustrates the fundamental fact that reform quickly unleashes forces that collide with the existing legal and ideological limits of the system. Led by the inexorable force of their own interest, managers and bureaucrats are bound to conflict soon after they have seemingly arrived at a *modus vivendi*. Technocrats are thus "objectively" (disregarding their protestation to the contrary) allies of capitalists—a fact that bureaucracy is fond of emphasizing.

Of course, one could suppose that some intermediate solution might be achieved: one that would allow managers to retain their autonomy, without introducing different capitalistic forms of property, and bureaucrats to preserve their political power. We recognize here the blueprint of the reformists—which is supposed to combine economic progress with the rule of the Communist party. But we must be aware that this equilibrium solution is contingent on continuous self-restraint by the technocrats, and on the good will of the bureaucracy. The former must constantly keep their "economistic" tendencies in check and observe the limits imposed by the system. The bureaucrats must genuinely accept segmentation of power as a permanent feature, and not as a short-run expedient to stabilize the economy. We must also not forget that an intrinsically totalitarian party bureaucracy is uncomfortable with such an arrangement and stands ready to recapture economic power as soon as the economy improves sufficiently.

The problems of transition are magnified if reform is accompanied by an opening toward the outside. Contacts between capitalist and socialist firms and their managers will make the wider scope of opportunities enjoyed by the managers in capitalism more apparent to their socialist colleagues. While there are no legal impediments for capitalist firms to use some of the efficient practices of socialist firms or to experiment with new methods of organization, whether in raising capital or in the management of the firm (worker participation, profit sharing, etc.), managers of the public corporations in (reformed) socialism will permanently stumble against legal (ideologically inspired) constraints on the firm's policy. This must, of course, increase the managers' dissatisfaction and make the equilibrium solution discussed in the previous paragraph even more precarious. Further, once free circulation of information—without which there can be no opening—is accepted, there is no reason why this should not extend to people and to capital. Once foreign capital comes in, new forms of economic life will exist in direct competition with the old. We thus again find that the natural drift of the reform is to push the frontiers of the permissible, introducing the elements of a mixed economy.*

The drift toward a mixed economy bares the question of the role of the party. The party has a role only if there is some economic power to go with it, that is, only if party functionaries can decide on nominations of managers and higher staff, on investments, types of production and so forth. Only thus does the party appear needed since without it economic life would ostensibly come to a halt; the other side of the coin is, of course, that by regulating economic life the party holds the levers of real power. When economic decisions are truly decentralized, the party apparatus becomes superfluous. The income of the middle-level bureaucracy may remain at the previous levels or even increase. But it is certain that in a reformed socialism their wages will be overtaken by the wages of successful entrepreneurs, managers, doctors, engineers, innovators.[21] Their relative economic position slips, while at the same time their real power (to make economic decisions) disappears. But the most dangerous development for them is that the reform opens up the question of the very *raison d'être* of the party. If the party becomes a debating society on the issues of ideology (far from economics) it could be innocuous enough. But why would such a party exist at all? The manager of a firm may not be bothered if the party organization in the local town convenes to discuss the latest ideological issues, as long as it does not interfere in economics. However, it is likely that sooner or later he would wonder what the real function of the party apparatus was. When such questioning becomes widespread, there is only one step to go to the extinction of the party as a ruling body. In reality, by abdicating its economic role it had already ceased to be a ruling party and its final demise would only be the logical conclusion of the process.

Only by being active in economic life can the party preserve its power. And by being active it impedes economic development. This encapsulates the dilemma of the reform. If the party wants to promote economic development, it must leave the economic arena. And by

---

*If it is true that the natural outcome of the reform is some form of capitalism, it raises a pertinent issue as to whether a dominantly capitalist economy may not be a "natural" economic system at the present level of development. The socialist blueprint may then be viewed as an attempt to reimpose an organic social structure on a society that has become individualistic. Put differently, if the development of productive forces leads to the emergence of an atomistic society, which in turn is compatible with economic growth, then an attempt to impose a socialist legal framework must stifle the development.

leaving it, it disappears from the scene altogether.

There are, it would seem, two ways in which the opposition of bureaucracy can be weakened sufficiently to allow the reform to proceed. The first—which may be called the Chinese way—consists in gradually changing the composition of the party constituency from bureaucrats to technocrats. Instead of top regional party bosses sitting in the Central Committee and/or managing the firms by virtue of their political position, the equation may be reversed: state firms' managers may become members of the Central Committee by virtue of their *economic* role. The second—tentatively called Hungarian—way attempts to edge the party apparatus out of economic decision-making. The party functionaries (i.e., professional party people) might sit in an upper chamber of parliament concerned with foreign policy, defense, and the general development of the country, while a freely elected lower chamber would be composed of citizens' or firms' representatives engaged in the daily nitty-gritty economic policy-making. This is similar to the approach adopted by the Jaruzelski regime in Poland: local-level democracy, along with development of the private sector, aims to create a social base for a regime that now almost entirely lacks one, and to transform Poland into a kind of socialist Spain of the Franco era. It would combine tolerable economic freedom, local-level democracy, and unchallenged authoritarian rule at the center. A movement from the totalitarianism of the Stalinist period to such a new social compact—an evolution charted in the last thirty years—is indeed progress. It is especially so, if we recall that in such authoritarian regimes a relatively liberal economic sphere in the long run tends to "spill" over into the politics and to lead to the transformation of the regime into a democracy.

So we reach the crucial question: could the Communist system survive the ascendance of technocrats and the unification of pragmatic elites? If the Communist party is transformed into a party of technocrats, the ultimate result would be a mixed economy and a multi-party democracy, an evolution similar to the one of Spain or South Korea. If the bureaucracy retains power, political immobilism and economic stagnation will be the lot of the socialist countries, and the international elites will remain split as they are now.

In other words, for the Communist countries now the issue is the same as the one faced by Russia in the late nineteenth century. To quote Hobsbawm (1979, p. 275): "As for the rulers of conservative, anti-bourgeois and anti-liberal regimes in Europe . . . they recognized, however reluctantly, that the alternative to capitalist economic development was backwardness, and consequent weakness. Their problem was how to foster capitalism . . . without also acquiring bourgeois liberal political regimes.'' With hindsight we know what the solution was. Rejecting capitalism, the aristocratic regime was overthrown by the very antidote to capitalism spawned by the capitalist system itself. And for that moment, it seemed that Russia was ahead in social, if not yet in economic, development. It had already reached the stage to which the West was supposed to accede only in the future. Yet today this appears to have been a detour, and the century-old problem resurfaces. What lessons does history hold? Will the historic meeting of Russia and the West finally take place, as it did for Germany in the nineteenth century, and Japan in the twentieth; or will the solution again be to embrace whatever negation of itself capitalism is currently generating; or finally, will the country remain in the morass of its present regime?

7.5 Prospects for Peace

Although the advent of technocrats and victory of the reform movement in the Communist countries face formidable obstacles, economic reality tends, in the long run, to impose itself. The opting-out (autarkic) solution is certainly feasible for some countries, but in a world of quick change and unstoppable communications it will become more difficult for such regimes to hold out for long. For some time, the combination of fervent nationalism, dictatorship, and police controls can stifle change and effectively insulate a country. Such solutions, as shown in the examples of Albania, Romania, or Vietnam, are effective for a period, but even then economic forces cannot be entirely suppressed, and the longer the view we take, the more important they become.

We saw above that a shared world outlook did not prevent World War I, any less than it made sixteenth- and seventeenth-century Europe peaceful. What are the prospects for world peace in the case of the advent of a technocratic elite in both the East and the West? Two powerful factors reinforce the likelihood of peace. The first is the sharing of a common econotechnic view of the world. The second is the existence of weapons capable of causing mutual destruction of belligerents.

Although a common *Weltanschauung* is more conducive to peace than mutually exclusive ideologies, it alone is not sufficient. Objective differences in the interests of countries and the people who rule them will always exist, and no level of friendship or "communality" will obliterate them. What is important is the field on which these differences are fought out. As naturally as the rulers of Europe in precapitalist times held that these differences were to find their expression on the battlefield, present-day technocrats hold that different interests are to be adjudicated by economic competition. For peace among nations, this is a momentous change. We can thus expect that econotechnicians of different countries will compete against each other (sometimes quite brutally), but that the competition will not spill over into the military arena.[22] Only rarely might it do so, but only when the forces of the adversaries are so lopsided that the victory of one is assured in advance. The invasions of Czechoslovakia and Grenada provide cases in point. However distressing such conflicts may be for the people affected, and particularly for those who are defeated, they do not represent a threat to international peace. They could pose such a threat only if one of the major powers were to engage in a string of small conflicts. This could be interpreted by other major countries as an attempt to attain international mastery by military means, and would be confronted by the same means. But our assumption is that the technocratic elite will strive to win mastery through economic competition, and may accept limited wars only if they are quickly winnable. And, after the experiences of Vietnam and Afghanistan, one might expect an even more cautious stance on the part of the technocracy.

The present relationship between the United States and Japan provides an example of the relations that might exist between countries ruled by technocratic elites. Former implacable foes in war have become implacable competitors in economics. The fact that they are political allies is immaterial, because the alliance is based on the existence of an external (Soviet) threat to Japan and (indirectly) to the United States. In effect, were that threat to disappear, the fundamental nature of Japanese–U.S. relations would not change: their economic competition is unlikely to slip into a military confrontation.

Now, if our view of the probable type of relations that will exist among econotechnic societies is correct, greater integration of the Soviet Union into the world economy is a welcome development. The greater the stake of the Soviet Union in the world order, the stronger will be the position of the reformers in that country, and also the less likely the outbreak of a generalized conflict. But Soviet integration into the world economy requires the type of people and ways of thinking prevalent in the world. The USSR needs such people, and not party ideologues, to further its interests in matters of economic cooperation with other countries.[23]

The second element that promotes peace—the existence of weapons of mass destruction—is obvious. Their role as a deterrent to conflict has been extensively discussed. It is clear that no government with anything like a responsible attitude is likely to provoke a war under these circumstances. A technocratic government would be even less likely to do so. The absence of a major conflagration in almost half a century since the invention of atomic weapons makes one confident that a generalized war (which would of necessity be a war of mass destruction) is difficult to imagine.

To sum up, the existence of important ideological differences dividing the ruling elites of different countries increases the likelihood of a conflict. The sharing of the same world view, consequently, reduces it. This does not imply that the countries' objectives must coincide, but only that the field in which these objectives are fought out, and the rules by which the conflict is adjudicated, are mutually accepted. A common view of the world, however, is not sufficient to prevent the outbreak of a conflict if the objectives of the elites are militaro-political (such as aggrandizement of national territory). The natural arena where these objectives must be confronted is the military one. But if the objectives are socio-economic, competition will tend to become economic, and political domination will be pursued through economic means. To paraphrase Clausewitz, economics is war pursued by other means. The tendency for a peaceful economic confrontation of the major powers will be reinforced by the existence of weapons of mass destruction which, in the case of a generalized war, would make the annihilation of the belligerents assured.

Thus, the advent of technocratic elites in both the East and the West would make the prospects for economic development in peace better than ever. But are we not thus underestimating the ideological and even irrational side of Man?

Sources of Data

Table 1.4: Sources and notes

France: Dieter Bos, *Public Enterprise Economics: Theory and Applications*, Amsterdam; New York: North-Holland, 1986, p. 426. Public sector defined following the strict majority of shares rule. Public sector share in total gross value added excluding financial services, but including agriculture. Public sector share in total employment excluding agricultural, but including financial services.

West Germany: Dieter Bos, ibid., p. 427. Public sector defined following the majority principle. Share in total gross value added. Share in total employment.

United Kingdom: Dieter Bos, ibid., p. 430. Public sector = nationalized industries. Public sector share in GDP. Public sector share in total employed labor force outside government proper.

United States: Dieter Bos, ibid., p. 432. Public sector = government enterprises [commercial enterprises owned by the government whose "operating costs are at least to a substantial extent covered by the sale of goods and services to the public" (Definition of the U.S. Department of Commerce)]. Public sector share in GDP. Public sector share in total employment.

Italy: Dieter Bos, ibid., p. 429. Public sector share in gross value added outside of agriculture and financial services. Public sector share in total employment excluding agricultural, but including financial services.

Austria, Australia, Denmark, Netherlands, Spain, Greece, and Portugal: R. P. Short, "The Role of Public Enterprises: An International Statistical Comparison," in Robert Floyd, Clive Gray, and R. P. Short (eds.), *Public*

Enterprise in Mixed Economies: Some Macroeconomic Aspects, IMF, Washington, D.C. 1984. Output data refer to the share in GDP at factor cost (Table 1, pp. 115–122). Austria, Denmark, Netherlands: Employment data from Charles Lindblom, *Politics and Markets*, Basic Books, New York, 1977, p. 114.

Turkey: Calculated from the World Bank CEM, *Sustaining the Adjustment Program*, vol. 2. Public sector share in terms of GDP. Public sector share in total non-agricultural employment (for app. year 1979): from A. Choksi, *State Intervention in the Industrialization of Developing Countries: Selected Issue*, World Bank Staff Working Paper No. 341, Washington, July 1979, p. 17.

Sweden, Finland, Norway, Canada: Frederic François Marsal, *Le Dépérissement des Entreprises Publiques*, Calmann-Levy, Paris, 1973, p. 16. Quoted according to Charles Lindblom, *Politics and Markets*, Basic Books, 1977, p. 114. Public sector share in total employment. No date given.

New Zealand: "Adam Smith's Islands," *The Economist*, March 5, 1988, p. 74.

Table 1.5: Sources and notes

South Korea: Young C. Park, "Evaluating the Performance of Korea's Government Invested Enterprises," *Finance and Development*, June 1987. Public sector share in terms of total employment and GDP.

Singapore: "Privatisation: Everybody's Doing It, Differently," *The Economist*, December 21, 1985. Public share in terms of GDP.

India and South Korea for 1971–72 from Leroy P. Jones, "Economic Factors in Public Enterprise" in Leroy P. Jones (ed), *Public Enterprise in Less-developed Countries*, Cambridge University Press, Cambridge, 1982, Table 2.1, p. 22. Employment for India from I. Shilin and S. Cheema, *The Public Sector in Developing Countries*, Progress Publishers, Moscow, 1987, p. 70.

Chile: Mary Shirley, "The Experience of Privatization," *Finance and Development*, September 1988, p. 34.

Niger and Sudan: Elliot Berg and Mary Shirley, Divestiture in Developing Countries, *World Bank Discussion Paper No. 11*, Washington, D.C., 1987.

For all the others: R. P. Short, "The Role of Public Enterprises: An International Statistical Comparison," in Robert Floyd, Clive Gray, and R. P. Short (eds.), *Public Enterprise in Mixed Economies: Some Macroeconomic Aspects*, IMF, Washington, D.C. 1984. Output data refer to the share in GDP at factor cost (Table 1, pp. 115–122). Employment data from footnote, p. 143.

Table 1.6: Sources

Italy and France: Saul Estrin, "The Role of Producer Cooperatives in Employment Creation," *Economic Analysis and Workers' Management*, No. 4, 1985.

United Kingdom: For number of firms: Co-operative Development Agency, *A Directory of Industrial and Service Co-operatives*, June 1984; for employment: The Labour Party's, *Labour's Charter for Co-ops*, 1985. Both quoted from Neil Carter, "Reports and Surveys," mimeo, pp. 182–187.

Spain, Belgium, West Germany, Netherlands, Sweden: William Bartlett and Phil Hobbs, "Sharing in Recovery: Co-operatives, Small Firms and Job Creation," mimeo, September 1987.

Table 1.7: Sources and notes

Hungary: Janos Kornai, "The Hungarian Reform Process: Visions, Hopes, and Reality," *Journal of Economic Literature*, December 1986, pp. 1687–1737. Employment = active income earners. Output = national income.

China: Edward K. Sah, "Major Contradictions in Peking's Economic Reforms," *Issues and Studies*, December 1986, p. 45. Output = gross industrial value added.

Poland: *Rocznik Statystyczny 1986*, Glowny Urzad Statystyczny, Warszawa, 1986, p. 61 (for employment) and p. 87 (for output). Also *Rocznik Statystyczny 1976*, p. 71. Employment includes agriculture. Output = national income (including depreciation). State sector includes cooperatives. Shares calculated at current prices.

East Germany: Anders Aslund, *Private Enterprise in Eastern Europe*, Macmillan, London, 1985, Table B4, p. 247 (for employment); Tables B6 and B7, pp. 249–250 (for output). Employment excludes agricultural employment. Output = net material product.

Czechoslovakia: *Statistical Yearbook 1987*, p. 135. State sectors includes cooperatives.

Soviet Union: *Narodnoe khoziaistvo SSSR v 1985*, Statistical yearbook, Finansy i statistika, Moscow, 1986. Estimation based on the share of agriculture in gross national output (around 16 percent) and the share of private production in agriculture (around 25 percent). All other "productive" activities are state-owned.

Annex to Table 1.9: Sources and Notes

Poland (employment in millions, 1985)

Capital	Labor	Entreprenurial role	Hired out	Total
Private	Entrepreneurial role	3.95*	0.93†	4.88
	Hired out	0	0	0
State	Entrepreneurial role	—	12.26‡	12.26
	Hired out	0	0	0
Total		*3.95*	*13.19*	*17.14*

Source: Rocznik Statystyczny 1986, Main Statistical Office, Warsaw, 1986, p. 61.

*Total employment in private agriculture (data for 1984).
†Total nonagricultural private employment (including owners).
‡All state-sector workers and cooperative-sector employment in agriculture.

United Kingdom (employment in millions, 1985)

Capital	Labor	Entreprenurial role	Hired out	Total
Private	Entrepreneurial role	2.5*	15.3	17.8
	Hired out	0	0	0
State	Entrepreneurial role	—	5.4†	5.4
	Hired out	0	1.3‡	1.3
Total		*2.5*	*22.0*	*24.5*

Source: Social Trends 17, Central Statistical Office, 1987 edition, London, Tables 4.7 and 4.9, pp. 73–74. Same source for 1971 and 1981 data.

*Total self-employed.
†Central government and local authorities.
‡Public corporations.

NOTES

Notes to Chapter 1

1. As the Yugoslav jurist Andrija Gams writes (1987, p.67; my translation): "Since the producer is considered to be an instrument of production—in Roman law the slave is the object, not the subject of law—the role of labor as a [distinct] factor of production is overlooked. Roman law therefore failed to develop terminology that refers to labor. For example, the labor contract (*locacio conductio operarum*) is regarded as a subspecies of a general hiring contract. . . . The work of a slave is viewed as the fruit of a thing (i.e., of a slave), and it is quite plain, according to the logic of the system, that it belongs to the slave owner—for the same reason that a fruit belongs to the owner of the thing."

2. Although the degree of right over workers is attenuated in feudalism as compared to slavery, there is no difference as regards the absence of the legal freedom of labor.

3. We do not include other concentration-camp–like systems either because their size was much smaller, or because they, unlike the Gulag, did not have an economic function.

4. The terminology often used in the past was *capitalist* and *rentier*. The first owned the capital and organized the production; the second merely owned (often loaned) the capital. It is not an adequate terminology for the present, because (at least since Schumpeter) there is recognition of a distinct factor of production—called the *entrepreneur*—who may own no assets. It is therefore more accurate to distinguish between (pure) capitalist and (pure) entrepreneur. Obviously, the two may be combined in one person—in which case we have a type that classical political economists called capitalist.

5. The same is true for bond-holders: they, as capital owners, also exert an indirect influence on management by refusing to buy new bonds or selling the ones they have.

6. Also, a pure public corporation (with 100 percent state-owned capital) will not be quoted on a stock exchange, so that timely information about the performance of the

firm will be unavailable. And it is precisely this information, in the form of stock market prices, that keeps the management of a joint-stock company on its toes. As Nuti (1987, p. 9) writes, the function of the stock market "is . . . bringing the current valuation of an enterprise as a going concern close to the maximum value, net of liabilities, that the enterprise productive assets could have if redeployed elsewhere in the economy or employed in the same activity under a different management and/or policy. . . . This function . . . is very important for bringing managerial capitalism somewhat closer to the traditional capitalist model in spite of the separation of ownership and control."

7. To quote Braudel (1984, p. 128): "Venice succeeded in establishing a system which from the very first raised all the problems of the relation between Capital, Labor and the State, relations which would increasingly come to be identified with the word *capitalism* in the course of its long subsequent development."

8. Examples of state-socialist firms are enterprises in centrally planned economies or, for example, industrial and commercial undertakings (*régies*) in France. In either case they lack a corporate (legal) personality although they may have independent accounting. In the Soviet Union firms are merely departments of the overall national economy. *Régies* cannot raise capital publicly, and are not quoted on the stock exchange.

Distinctions among different levels of state intervention in the economy are clearly reflected in Francophone countries in the names of companies. (1) *Sociétés à participation d'état* are simply private capitalist firms in which the state owns some shares. These firms are not different from other capitalist firms. (2) *Sociétés d'économie mixte* are in effect public corporations: the state is the predominant shareholder (often more than 50 percent of capital), but the autonomy of the board of directors is preserved. They can raise capital on commercial terms and be quoted on the stock market. (3) *Sociétés d'état* are wholly state-owned (e.g., in France, the Bank of France, nationalized banks and insurance companies). Like *Sociétés d'économie mixte* they are subject to the ordinary commercial law, but may often have an *agent comptable* who is the representative of the government and who decides on expenditures. To some extent their independent decision-making is restricted. (4) *Établissements publics* are also entirely owned by the state (e.g., French electricity, coal mines, radio and television, railroads). They have a corporate personality and are generally subject to commercial law; yet the state role is more strongly exercised and their position is basically the same as that of state socialist firms. (5) Finally, *régies*, as mentioned before, are direct state-managed industrial or commercial undertakings (e.g., PTT services, *Renault*). Forms (4) and (5) belong to the state-socialist type; forms (2) and (3) to the public-corporation type.

A similar situation exists in South Korea: terminological differences reflect differences in the degree of state control. Type (1) is called government-backed enterprise (less than 50 percent of state ownership). Types (2) and (3) are government-invested enterprises (more than 50, and up to 100 percent of state equity; the state also appoints the top manager; examples include Korea Development Bank and National Power Company). Finally, type (5) are called government enterprises—they are in effect government departments. (From Young C. Park, 1987.)

9. The term "public corporation" should be evocative of the nature of such firms. "Public" means that the state has entire or controlling interest in the enterprise; "corporation" (or "company") indicates its commercial nature. It should be noted that some real-life public corporations (e.g., corporations established by statute for specific purposes) may belong to (in our terminology) the state-socialist rather than public-corporation type.

10. An interesting case are ESOP (Employee Stock Ownership Plan) firms in the

United States. The position of the management there is similar to the one in a capitalist joint-stock firm—the entrepreneurial role is shared between owner-workers and managers. The difference with joint-stock companies is that workers and owners in ESOPs are the same people. Similarly, worker share-ownership schemes (there are around a thousand of these covering approximately 1 million workers in the UK) introduce some elements of the cooperative-1 type into the still predominantly capitalist firm.

11. This is important because, in a few cases, some of the central state organs' power was merely transferred (or lost) to regional bodies, without this having had a perceptible impact on the position of enterprises or on the nature of the *state*-planning system.

12. Abundant literature exists on the implication of different maximands for the reactions of the firms. This was spawned by interest in the economic theory of the labor-managed firm. The recent origin of this interest dates to a Benjamin Ward (1958) article; the realization that the contrast between the reactions of the entrepreneurial and the labor-managed firm is only a special case of the behavioral implications of different maximands, was first clearly spelled out in Dubravčić (1970). Yet in Marxian economics it was always very clear that differences in modes of production imply differences in relative prices: transition from values (labor prices valid in conditions of petty commodity production) to prices of production is exactly that. Wicksell (1970, pp. 120–21) was also aware of it. Later, in the wake of Walras, the concept of the entrepreneurial firm entirely took hold of economists and other types of firms began to be regarded as oddities. In a personal recollection, I was once asked to classify some articles following the *Journal of Economic Literature* classification scheme. I placed an article dealing with a labor-managed firm under the general *Microeconomics* heading. The reason was simple: everything that deals with reactions of an individual firm operating in a decentralized environment, regardless of its objective function, should be considered microeconomics. The article, however, was reclassified under the heading "Socialist economic systems" (where *pêle-mêle* were works on microeconomics, macroeconomics, economic history, etc.).

13. Commenting on Britain, John Westergaard and Henrietta Resler (1975, p. 212) write: "The [public] corporations never . . . asserted the role for themselves directed against the dominance of private capital. . . . This was clearly ruled out by the absence of any positive directives to them to take on such a role . . . and by their inevitable practical links in those circumstances with private industry, through the everyday conduct of business as well as through the recruitment of board members." And further: "it would be impracticable for public sector industry as a minority element to pursue investment, pricing and pay policies which would put it widely outside the general market conditions set by the predominance of private enterprise" (ibid. p. 215).

14. Note, however, that one part of entrepreneurial income in venture capital firms accrues to the supplier of capital, which is not the case in a theoretically pure entrepreneurial firm. But on the other hand, to the extent that capitalist firms gradually use more borrowed capital, they move closer to the entrepreneurial type; in the extreme case when the equity capital is zero (or shares do not carry the voting rights) and the borrowed capital is 100 percent, the firm becomes entrepreneurial.

15. From David M. Blau (1987, p. 447). Blau also notes that the change in trend occurs after more than a century of steady decline in the proportion of the self-employed. Moreover, the change is concurrent with similar developments in Japan and several West European countries.

16. From *Social Trends 17*, Central Statistical Office, London, 1987.

17. The exceptions to that rule are essentially the entrepreneurial firm and the public corporation, and only partly the joint-stock firm. Management

buy-outs further reduce these exceptions.

18. The best-known type of labor-managed firm is the Yugoslav type. Until 1971 the firms were obliged to pay a rent on capital to the state, and thus clearly belonged to the labor-managed type. After that date, however, the situation is ambiguous: firms (i.e., workers) appropriate all of the income, including income from capital, and are, in this respect, equivalent to the cooperative-1 type. Yet since they cannot freely dispose of the capital, one could also treat them as labor-managed firms that hire state-owned capital at zero rental.

19. This also translates as an equivalent decline in total employment because the overall number of the employed was constant.

20. Yet the ratio in 1985 was practically the same as in 1971. The structure, however, was different: in the private sector there was a shift toward self-employment; in the state sector there was a shift away from public-corporation employment and into state (essentially government) employment.

21. Since the writing of this text most statutory limits on the private sector in Poland have been removed.

Notes to Chapter 2

1. The liberation of peasants from noneconomic bonds and duties was not achieved in Europe, Russia, and the Americas (with a few exceptions) until the 1860s. In Ethiopia slavery endured until 1935 (and provided one of the reasons alleged by Italy for the invasion of the country) and in Kuwait until 1963.

2. See Janos Kornai (1986).

3. W. Friedmann (1974, pp. 370–81) defines the possible role of the state in the economic sphere in terms of the following functions: (1) state enterprise, (2) mixed public-private enterprise, (3) regulatory public controls over private enterprise (price and wage controls, protection from competition, taxation, control through licensing, currency and trading restrictions), (4) assistance to private industry, (5) managerial participation by government, (6) coordination of public and private enterprise in economic planning. Friedmann's points (1), (2) and (5) are concerned with the definition of different modes of production, and not with what we regard as state interference as such. Similarly, Friedmann's point (6), the coordination function, is not an independent function but derives from the existing structure of production; generally, the greater the proportion of the economy's assets owned by the state, the greater would be the tendency (and temptation) to resort to a centralized coordination of economic decisions. All regulatory controls (Friedmann's (3)), except taxation and protection from competition, are contained in our category (a) of government interference: limiting the range of choice of individual economic agents. Taxation represents our category (c). Friedmann's point (4) plus protection from competition are equivalent to our category (b).

4. We have in mind only productive agents, not consumers for whom choice remains, even under central planning.

5. In the labor-managed system, for example, where the state only collects interest from the return on capital, individual enterprises freely pursue their own objectives and the coordination mechanism is clearly decentralized. Strictly speaking, there is no reason why that system would have more government interference than a pure capitalist system.

6. The only exception is the state-socialist mode where the mode itself implies the level of state intervention (central planning).

7. By horizontal links we mean purchase, sale, credit, etc. relations into which

firms (as individual maximizing agents) freely enter with other firms (banks) in the economy.

8. It seems difficult to envisage other dominant modes of production based on private ownership of capital as realistic alternatives.

9. That is, by changing the legal framework so as to allow the development of the forms that have previously been legally restricted, as for example, allowing greater employment by capitalist enterprises, private ownership of land, or easier formation of cooperatives. In a capitalist system, only privatization (or nationalization) of assets involves such a change in the legal framework.

10. There is also no reason why they should compete as they all belong to the state and cannot keep any profits for themselves.

Notes to Chapter 3

1. It is essential to deduct disutility of labor from the utility of goods and services produced, in order to arrive at a correct estimate of well-being. The inadequacy of the current accounting system is particularly obvious in national accounts where wages are entirely included as part of the net product. As we do for capital with depreciation, we would need to deduct from wages that part which simply compensates for the expansion of labor during the process of production (in other words, that part of output which is needed to place the worker back on the same utility level at which he was before the beginning of the process of production). Note that Ricardo was essentially right in treating only profit as net product: if wages are, as in his time they indeed were, at the existential level, i.e., roughly equal to the "depreciation" component of labor, then only profit represents a surplus.

2. ". . . [taxation] is almost inevitably in the nature of an injury to the productive process" (Schumpeter, 1947, p. 198).

3. The difference in the accounting efficiency, reflecting the care with which the use of the resources and revenues is monitored, is displayed in the delay with which the double-entry bookkeeping was introduced in the management of state funds. As Schumpeter (1947, p. 123) writes, "capitalist practice turns the unit of money into a tool of rational cost-profit calculations, of which the towering monument is double-entry bookkeeping." And, further, "Luca Pacioli's treatise on bookkeeping, 1494, supplies by its date an important milestone. For the history and sociology of the state it is a vital fact to notice that rational bookkeeping did not intrude into the management of public funds until the eighteenth century and that even then it did so imperfectly and in the primitive form of 'cameralist' bookkeeping."

4. Note that the same problem does not exist with respect to the use of capital. Capitalists alone decide where they want to invest (in one firm or another) and alone bear the consequences of their choices.

5. We can thus expect to find that efficiency of investments in all modes of production with state ownership of capital is less than when capital is privately owned.

Notes to Chapter 4

1. These figures refer to the gross value added, which excludes financial services. Quoted from Dieter Bos (1986, p. 426).

2. Save for the industrial sector in Austria. Quoted from *Le Monde*, October 24, 1985.

3. Quoted from Roland Drago, "Public and Private Enterprise in France," in Wolfgang Friedmann (ed.) (1974).

4. We use the word "preference" because the fully fledged centralized state sector was not created in Western Europe.

5. In a book written in 1959 Calvin Hoover (1959, p. 300) writes: "The realization that the Soviet system was not a workers' state had developed slowly, but it had come to limit substantially the desire for any comprehensive program of nationalization [in the West]. The possible effect upon personal liberty of the management of an entire statized economy by a huge bureaucracy . . . had come to be a factor in setting the policies of the [British] Labour Party."

6. In effect, the marginalization of communist parties in Western Europe is already happening. French Communists, who in the late 1940s and early 1950s commanded 30 percent of the national vote, fell below 10 percent for the first time in a general election in 1986. In West Germany's first postwar election (1949) the Communists obtained about 6 percent of the vote; practically ever since, their share has been close to zero. Immediately after democratization in Spain, the Communists won around 10 percent of the vote; their share is now less than 5 percent. In Portugal Communists reached their peak in 1979 with 20 percent of the vote; they have steadily declined, down to 12 percent in the 1987 election. Italy is only apparently an exception. The share of the Communist vote there climbed through the sixties and the seventies to reach almost 35 percent in 1976; since then the Communists have lost 10 percentage points.

7. As Marx writes in *Critique of the Gotha Program*: "Any distribution whatever of the means of consumption is only a consequence of the distribution of the conditions of production themselves. The latter distribution, however, is a feature of the mode of production itself. The capitalist mode of production, for example, rests on the fact that the material conditions of production are in the hands of non-workers in the form of property in capital and land, while the masses are only owners of the personal condition of production, of labor power. If the elements of production are so distributed, then the present-day distribution of the means of consumption results automatically. If the material conditions of production are the co-operative property of the workers themselves, then there likewise results a distribution of the means of consumption different from the present. Vulgar socialism . . . has taken over from the bourgeois economists the consideration and treatment of distribution as independent of the mode of production and hence the presentation of socialism is turning principally on distribution. After the real relation has long been made clear, why retrogress again?" (In Robert C. Tucker, ed. (1978), pp. 531–532.)

8. With n economic agents (firms) the number of possible links is $n(n-1)/2$. Then, every time a new firm appears, the number of possible links increases by the factor of n.

9. The same idea about the lesser relative inferiority (or even superiority) of centralized planning when the organization of the economy is simple, lies behind the theory formulated by Branko Horvat (1982, pp. 202–204), according to which centralized planning represents an efficient instrument for effecting rapid modernization (equated with industrialization) of the economy. Yet as the economy develops, central planning becomes an obstacle to further transformation.

10. Oskar Lange defined the Stalinist command economy as a *sui generis* war economy. Quoted from Herbert Block, "The Economic Basis of Soviet Power," Appendix 1 in E. Luttwak (1983, p. 120).

11. See E. H. Carr (1966, pp. 98–100, also p. 91). Also E. Luttwak (1983, p. 22): "[Planning] not itself part of the Marxist inheritance, but rather the offspring of the systems of economic control invented in both Germany and Britain during the great war just ended (which made possible the huge and indeed improbable increases in war production of the two countries), the direction of the economy by central planning seemed an innovation of epic proportion to Soviet leaders."

Testimony to the strong intellectual influence of planning ideology at the time is found in the following review by Keynes (published in 1915) of a typical German work of that period. According to the author, Keynes writes, "even in peace industrial life must be mobilized. This is what he means by speaking of 'militarization of our industrial life.' . . . A system of regulations must be set up, the object of which is not the greater happiness of the individual . . . but the strengthening of the organized unity of the state for the object of attaining the maximum degree of efficiency. . . . In particular, the coming peace will bring with it a strengthening of the idea of the State action in industry. . . . Foreign investment, emigration, the industrial policy which in recent years had regarded the whole world as a market are too dangerous. The old order of industry . . . is based on Profit; and the new Germany . . . without consideration of Profit is to make an end to that system of Capitalism, which came over from England one hundred years ago." (Quoted from Hayek, 1944, p. 183.)

12. To realize how far Soviet prestige has sunk in Western Europe, consider this quote from Schumpeter (1947, p. 400), written in 1947: "They [many, not necessarily communist, Frenchmen] look upon Russia as 'the great fact of our day', the power that . . . really matters, the power to which *il faut s'accrocher* and with which, in order to be reborn, France must side, against England and the United States, in any future struggle."

13. What we mean is the following: Until the early 1960s it was difficult to be sure whether Romanian growth rates were or were not comparable to Japan's. But it is ludicrous to maintain now that the pace of development of the two countries after World War II was the same. In effect, as Table 4.1 illustrates, Romania's official average growth rate since 1950 is 8.6 percent p.a.; Japan's only 7.6 percent. Then clearly either the Romanian data must be forgeries, or the Romanians must be using a totally arbitrary weighting system in calculations of their national product.

Table 4.1

Japan's and Romania's Rates of GDP Growth (1950–1985)

Period	Japan	Romania
1950–55	8.5	13.9
1955–60	8.5	6.9
1960–65	12.4	9.0
1965–70	11.0	7.7
1970–75	4.3	11.3
1975–80	5.0	7.2
1980–85	3.9	4.2
1950–85	7.6	8.6

Sources: Japan: 1955–1985 from the International Monetary Fund (1986). For 1950–55, rate of growth of per capita national income, augmented by the population growth rate. Romania: Rate of growth of net material product; from *Monthly Bulletin of Statistics*, United Nations, and *Yearbook of National Accounts*, United Nations (various issues). For 1980–85 from the International Monetary Fund (1986).

14. In an interesting paper Abram Bergson (1987) compares four socialist (the USSR, Hungary, Poland, and Yugoslavia) with seven capitalist economies. He finds

that the output per worker in socialist economies (adjusted for the differences in the per-worker capital stock and farm land) is at least a quarter below that in the capitalist economies. Introducing the dummy variable for the social system (socialism), Bergson finds that in all cases it is significantly negative. Similarly, in an unpublished paper Katherine Terrell (1987) finds that in the decade of high investment growth in the 1970s, the productivity of the Western-imported capital in Poland was nil.

15. An illustration of the increasing gap between capitalist and state-socialist economies whose starting points were approximately the same is provided by studies that compare actual levels of standard of living in Czechoslovakia and Austria, and in Poland and Austria. Comparison of the actual levels of living standards (personal consumption) seems more appropriate than the more conventional comparison of GDPs per capita because it dispenses with problems arising from differences in price structures, accounting conventions, and GDP composition (e.g., high share of investment expenditures, which obviously does not help the current level of consumption and, in the case of low efficiency of investment, need not help it in the future either). Comparison of standards of living is also the real yardstick individuals use when they measure the relative attractiveness of different countries and economic systems.

On the basis of a joint study of consumption levels carried out by Austrian and Polish statistical offices, Askanas and Laski (1985) report that in 1964 Polish per capita private consumption amounted to 62 percent of Austria's (measured in Austrian prices as weights), 57 percent in 1973, and 55 percent in 1978. Similarly, Havlik (1985) calculates that in 1980, total (wage and "social"—the latter including government expenses for social purposes) per capita household income in Czechoslovakia was equal to 49.1 percent of Austria's level. In 1964, it amounted to 76.7 percent.

Eastern Europe's falling behind in respect to countries that were on a similar level of development is also illustrated by data that compare Czechoslovakia and Hungary with, respectively, Austria and Greece (Table 4.2). We see that in terms of availability of phones both Czechoslovakia's and Hungary's relative positions were worse in 1983 than in 1975. Similarly, Hungary's advantage over Greece with respect to television sets and access to medical care has slipped. In six out of eight possible time-wise comparisons, Czechoslovakia and Hungary are either increasingly falling behind Austria and Greece, or are seeing their advantage reduced.

Table 4.2

Comparison of Czechoslovakia and Hungary with Austria and Greece

	No. of inhabitants per doctor:			No. of phones per 1000 people		No. of TV sets	
	1975	1980	1983	1975	1983	1975	1983
Austria	540	397	441	281	421	253	298
Czechoslovakia	420	363	354	176	210	249	281
Hungary	389	347	330	99	121	218	262
Greece	490	411	394	221	302	116	160
Ratios:							
Czech./Austria	1.29	1.09	1.25	0.63	0.50	0.98	0.94
Hungary/Greece	1.26	1.18	1.19	0.45	0.40	1.88	1.64

Source: Stajner (1986).

*The data are presented so that a number greater than 1 indicates relative advantage.

16. They were, by and large, what Kornai calls "naive reformers." To quote Kornai (1986a, pp. 1728–29), "the faith placed in the harmonious, mutually correcting duality of 'plan' and 'market' . . . is the centerpiece of the pioneers' naiveté." They failed to "foresee many complications which, as it turned out, are the barriers to consistent application of their proposals."

Notes to Chapter 5

1. Quoted from Kornai (1986a, p.1709).

2. According to *The Economist* ("China's Economy," 1 August 1987), 570 enterprises in Shanghai alone have been leased out in 1987. They are all small-scale enterprises, the largest employing 400 people.

3. This number is 9 to 12 in Hungary (raised to 500 in 1988), 10 in East Germany and Yugoslavia. The only exception is China and, from 1989, Poland. In Chinese terminology, "individual" sector includes all private firms with less than 7 workers. Capitalist sector includes all firms with more than 7 employees, so there is no effective ceiling on hiring by private firms. The two sectors employ more than 23 million people (20 million for the individual and 3.6 million for the capitalist). The 1987 Individual Labor Law in the Soviet Union which allowed private enterprises with *no* hired labor offers a good example of the difference between owner-worker and capitalist firm. Similarly, the Soviet Cooperative Law of June 1988 permits only cooperative-1 type: there must be neither hired labor, nor hired capital, i.e., capital belonging to non-members. By mid–1988, total employment in the cooperative and "individual" sector amounted to some 620,000 people (quoted from *Le Monde*, August 4, 1988, p. 3).

4. Since pure public corporations hardly exist in socialist countries they are subsumed under state-socialist firms. In other words, all firms that are somewhere in-between state-socialist and public corporation still share more characteristics with the former type.

5. The position of top regional leaders is ambiguous. By their importance they clearly belong to the high-level bureaucracy. Yet by their objective position—the criterion which we use to distinguish between the high- and medium-level bureaucracy (i.e., central vs. regional responsibilities)—they belong to the middle-level bureaucracy.

6. It is certainly much more correct to speak of state planning rather than of a centrally planned system in China. The former also includes planning and allocation of outputs and inputs by provincial and county organs, which is indeed a much more prevalent practice in China. A similar devolution from the center to the republican level was aptly dubbed in Yugoslavia "polycentric statism."

7. Strictly speaking, state orders in Poland are not legally binding. Enterprises can theoretically reject them. This possibility is, however, only of a theoretical nature because the government disposes of fairly strong inducements. It guarantees the supply of raw materials (which are generally in short supply), and may further prevent an enterprise which has rejected a government order from (say) exporting its output. Further, if inducements fail, an enterprise may be forced to accept government orders by its "founding organ" (generally a ministry or local authority). In the Soviet Union, state orders are even legally compulsory.

8. Tha data for 1985 refer to the first half of the year [from D.H. Perkins (1988, p. 614)].

9. Primarily because the transition would involve loss of jobs for a number of them, tighter work discipline, and increased wage differentials (to be opposed by the unskilled workers).

10. See Josip Županov in the Yugoslav weekly *NIN*, January 30, 1987.

11. For a comparison of the position of Polish and Hungarian cooperatives see Szul and Kirejczyk (1987) and Kornai (1986a, pp. 1701-1702).

12. During the 1958-1979 period the land in China was effectively owned by the state. The communes were subject to compulsory deliveries to the state, functioned within the framework of central planning, and the role of individual incentives for communes and its members was all but nonexistent. *De facto* privatization of land started in 1978-79. The land is still officially owned by the state but it is leased free of charge for 15 years (and, in some places even longer) to private cultivators. Until 1984, farmers were obliged to sell at government prices a prescribed quantity of output to the state; after that date they sign contracts with the state which are only nominally voluntary. From 1988 the land is transferable. The percentage of agricultural households covered by the "responsibility system" increased from 29 percent at the end of 1979 to 98 percent in 1983. Quoted according to Sah (1986) and Aubert (1984, p. 3).

13. For example, in 1986 the private plots in Soviet Union, which represented 1.6 percent of total cultivated area, accounted for 27 percent of total egg and meat production, 21 percent of milk, and 25 percent of wool production. (Quoted from *The USSR in Figures for 1986*, Central Statistical Board of the USSR, Moscow, 1987). In Hungary, private household farming contributed in 1984 about a third of total agricultural output (Kornai [1986a, Table 4]). Finally, between 1978, when the communal system was disbanded in China, and 1983 Chinese grain production increased by 23 percent, cotton production by 114 percent, edible oils by 102 percent, and meat by 64 percent (from Aubert [1984]).

14. Even if it is not quite clear what, in this context, the market price is. For this opens the whole new issue of valuation of shares, i.e., of the capital market.

15. Exactly this was done by the new Polish law on economic activity: the ceiling on private-sector employment as well as for joint ventures between the private and state sector was lifted altogether, while all but eight industries became open to the private sector.

16. As a Chinese commentator puts it: ". . . it is also a fact that the difference between these [private sector] incomes and those made by employees in the state owned businesses is extraordinary. If we allow things to go on this way unchecked, we will discourage not only enthusiasm of employees in the state-owned businesses but also the initiative of self-employed people for production and operation, reducing their progressiveness and increasing their parasitism." (Tao Youzhi [1987, p. 41].)

17. They have an interest in keeping the entry limited and rules murky. It is only under such nontransparent rules that their connections with the bureaucracy can pay off. For example, in Belgrade, easier access to taxi licenses for newcomers was opposed by the association of private taxi drivers under the pretext that a greater number of taxis endangers traffic safety. One is reminded of Bastiat's appeal of candlemakers against the disloyal competition of the Sun.

18. The same objection applies to the sale of shares to workers since this is, as we saw above, likely to result in usual "outside" privatization. The situation, however, is somewhat different for the small-scale capitalist firms where the owner directly participates in the process of production. They can be considered remnants of the past, hailing from the previous "socio-economic formations" (like petty commodity production) and thus to be by historical necessity bound to disappear. It is much more difficult to argue the same for a capitalist joint-stock firm, the hallmark of the capitalist organization.

19. And firms may be interested in it at all only when the budget constraint becomes harder. So long as their investments are financed from the government's grants or soft credits, they will be understandably indifferent toward the possibility of issuing bonds.

20. China's so-called stock market is in reality a bond market. Certificates specify a guaranteed as well as the maximum return, and they carry no management rights.

21. Tax revenues, of course, can go up if lower tax rates are more than compensated by an increase in profits.

22. This might happen: in order to offset more liberal policy in the areas of licensing and physical constraints on growth, the state may decide to increase taxes.

23. On data for Czechoslovakia (Table 5.6) one can observe the trend in historical perspective. Already before Communists' seizure of power (February 1948) Czechoslovakia disposed of a high share of state-owned industry: most of it was nationalized after the war by the Social Democratic government. Following the February 1948 coup, the state sector expanded to cover about three-fifths of the national economy. By 1960, the private sector was practically eliminated, and the state had absorbed most of the cooperative sector.

Table 5.6

Czechoslovakia: The Structure of Production (in % of national income)

	February 1948	After February 1948	1960
State sector	50	61	96
Owner-worker	25	25	4
Capitalist sector	25	14	0
Privately owned capital	50	39	4

Source: Blaha (1986, p. 140).

24. The difference in the mental framework between technocrats and bureaucrats is seized by Aron (1964, pp.177–78): "Il y a, d'une part, les hommes du parti, et de l'autre les techniciens. Ceux-ci voudraient être rationnels à la maniere de l'inspecteur des Finances françaises, et ceux-la sont soucieux d'idéologie ou de l'opinion des masses." Note also that both Veblen's engineers and economists would fall into the category of technocrats.

25. For a compact and readable exposition of the Kornai-type mechanism see "The Reproduction of Shortage" and "Hard and Soft Budget Constraint" in Kornai (1986). The idea of the soft budget constraint is present already in Mises (1963, p. 709).

26. The system can be described as one of "supplications and threats." Supplication of all actors to deliver promised goods, to give loans, etc. Threats that if they fail to do so, they will be reported to higher authorities, will have to bear the responsibility for the fall in production, etc.

27. The costs cannot be so high that the regional bureaucracy gets itself into trouble with the higher-ups for failing to reach some planned level of production. If only a potential increase in production is forgone, who is ever to know this?

28. Note that these conditions for the power of bureaucracy are not limited to socialist countries. A capitalist market economy in which bureaucracy is powerful (e.g., Argentina) similarly fulfills the same conditions (with a possible exception of financial indiscipline—and this outside the state sector only). It is a subject for historians to study how introduction of modern civil laws only provides a screen behind

which actual lawlessness reigns. In the Ottoman Empire, for example, written law did not exist. Some present-day societies only externally resemble legal states, while in reality they are no different from the Ottomans. It is somewhat ironic that the existence of modern law with its discretionary application represents a condition for a genuine legal anarchy.

29. Although economic growth (with rising wages) and the market economy are, we believe, correlated, the immediate effects of the market economy on the level of real wages would probably be negative, particularly if one takes into account that reforms are most likely to be implemented in conditions of (1) economic crisis when austerity rather than growth is probable, (2) a history of financial repression which in order to be corrected (i.e., firms and households to be induced to save again) will necesssitate a transitory period of relatively high real interest rates.

30. It could be argued that workers, unlike bureaucrats or technocrats, do not dispose of institutional channels to influence the outcome of the debate, nor do they participate in the decision making. They have, however, other no less important means. If introduction of reforms is not accompanied by strikes, big increases in unemployment, or workers' passive resistance to tighter work discipline, but results in a fast increase of production and, even more importantly, creates a perception that there is a clear correlation between the work effort and reward, forces of reform will obviously be strengthened. And *mutatis mutandis.*

31. Consider the advice given by Schumpeter (1947, p. 221) to central planners in a Lange-Lerner world: ". . . we will simplify the transitional problems [transition from capitalism to socialism] before the new ministry or central board . . . by assuming that they will leave farmers substantially alone. This will not only eliminate a difficulty that might well prove fatal—for nowhere else is the propery interest so alive as it is among farmers or peasants . . . but also bring additional support, for nobody hates large-scale industry and the specifically capitalist interest as much as the farmer does.''

32. Not only because their relative level of income is higher but also because they receive other benefits more liberally: free health care, earlier retirement, longer maternity leave.

33. It may be interesting to note that although peasants working on their own land and owner-workers (the so called ''individual sector'' in socialism) belong to the same mode of production, their position with respect to the reform is different. ''Individual sector'' people are in favor of it; peasants are neutral. This is explained by historically different levels of political involvement between city and rural dwellers, and greater security of private property (of land) felt by the peasants.

34. If the land is privately owned and most peasants slip into neutrality, the prospects for the pro-reform movement are weaker: it must win over all workers in the state sector or a majority of the high-level bureaucracy, neither of which is easy.

35. This was most dramatically illustrated by the negotiations in Gdansk, Stettin, and Katowice when the August agreements creating ''Solidarity'' were signed between direct representatives of the workers and top-level state bureaucrats.

36. Of course, the conservative role of the peasantry has been noted by many. A typical nineteenth-century view is quoted by Eric Hobsbawm (1979, p.199): ''The peasantry has at all times been the most conservative element of the state. . . . Its appreciation of property, its love of native soil makes it into the natural enemy of urban revolutionary ideas and a firm bulwark against social-democratic efforts. It has therefore been rightly described as the firmest pillar of every sane state.''

37. To the extent that the October Revolution was a workers' revolution at all, which, of course, is beyond the point we discuss here.

38. This is apparent in the Gdansk and Stettin agreements. Out of respectively 21 and 28 points about half (exactly 12 and 13) are essentially syndicalist. They include

things like creation of free trade unions, increase in wages, pensions, travel and family allowances, maternity leave, etc. An analysis of the agreements shows that the demands formulated by workers can be broadly divided in three categories. Other than syndicalist, there are general economic demands (e.g., increased supply of foodstuffs, introduction of rationing cards, elimination of sale of Polish products for hard currency) and political demands (freedom of speech, liberation of political prisoners, elimination of the privileges enjoyed by the party apparatus, police and the security services, improvement in relations with the Church). For the texts of the Gdansk, Stettin, and Katowice agreements see Oliver MacDonald, ed. (1981).

39. In the last years of Stalin's rule this became fairly obvious. His paranoid character and extreme suspiciousness meant that no one was safe. Sometimes, the closer one was to power, the more was he suspect in Stalin's eyes and the more was his life threatened. Only Stalin's death prevented the planned purge of the Central Committee and a probable replay of the great terror years 1936–1939. But in this case Stalin's paranoia would have made murders more random. If such a senseless destruction of the "apparat" had taken place, there is little doubt that the system would have been seriously weakened. The longer had such an orgy of terror continued, the more likely would have been a reversal after Stalin's death. (With some irony it could be said that Stalin by his early death helped save the key repressive features of his system.) This also explains why the ten years of chaos of the Chinese Cultural Revolution precipitated the emergence of the reform movement.

Notes to Chapter 6

1. In Marx's terminology: petty commodity production.

2. Closely related to this is the underdevelopment of stock and financial markets in LDCs.

3. See the sectoral distribution by countries in R. P. Short (1984, p. 124 ff.).

4. For the same reason no capitalist (or, in general, private) firm will operate in these sectors.

5. The difference will be in environment. The state-socialist firm in capitalism will function in a milieu where decentralized coordination predominates.

6. The political elite in developed (democratic) capitalism includes all professional party politicians. One group of politicians (belonging to the party or parties in power) holds the levers of state power. They represent the top of the executive branch of government. Another part of the political bureaucracy are people who form (1) the legislative branch of the government, viz., representatives at different administrative levels—federal, provincial, communal—as long as politics is their main occupation, and (2) professional party politicians working in the standing organs of political parties. The dividing line between political bureaucracy and the state administrative apparatus will not always be clear. As politicians we may consider all those whose "fortunes" depend on the current constellation of power among political parties. They are the people who are affected by the outcome of the political (i.e., electoral) process. On the other hand, those whose role is more technical, and whose position is generally little changed if one party replaces another in the government, belong to the state apparatus.

7. There are obviously people in socialism whose views (like those of blacks, for the Conservatives, but not for Labour) are beyond the pale of the admissible. These are, for example, people who would question socialism's political set-up, or the dominant (state) ownership of the means of production. They are doubly oppressed: they are not allowed to present their views, are not represented in the ruling party, and cannot found another party because its existence is not permitted. In that sense, the difference between political democracy and a dictatorship is that in a

dictatorship some groups are virtually excluded from the polis.

8. In an interesting article in the West German newspaper *Die Zeit* (reported by *Ekonomska politika*, September 7, 1987, p. 41) all employees in West Germany are divided into several groups according to the job security and fringe benefits they enjoy. Government administration and state-sector employees (in railways, post office, insurance), i.e., about 16 percent of the total active population, belong to what the author calls the first—most privileged—caste. Their jobs are "practically guaranteed, wages certain, and retirement benefits particularly well insured." To the second group (3 percent) belong employees in public enterprises, research institutions, some mass media, etc.—undertakings not free from government influence. The third group (13 percent) is composed of employees in big private firms protected by the government. The numerically dominant fourth group represents the "hard core" private sector: those working in medium and small firms and the self-employed. They account for about 60 percent of the total active population. The last group are the unemployed, i.e., more exactly, those with a sporadic employment record (8 percent of the total). It is significant that the likelihood of being fired is 7 times higher in private firms with less than 20 employees (fourth group) than in firms employing more than 1000 (roughly, the third group).

9. The same is true for the state-sector managers in capitalism. Due to their small numerical and political importance, they cannot represent a separate social group. We may subsume them under the state administration.

10. But not when it comes to the distribution of the net product.

11. Coincidence of interests is evident even from a casual perusal of the press. Suffice it to note that in all examples involving the restructuring of the economy, workers, capitalists, and managers from the declining industries jointly clamor for government support: steel and cars in the United States, textile and coal in Western Europe. And obviously the support they derive can only be at the expense of some other workers and capitalists at home.

12. In the extreme case, of course, elimination of all tariffs and tariff-like measures will satisfy all the criteria: liberalization, equal treatment of all, and probably economic efficiency.

13. And thus a covert subsidy.

14. This is the case in British Telecom where employees own a sizeable share of the total stock. It is estimated that in the United Kingdom 1.5 million workers (about 8 percent of the total private-sector employment) own shares in companies where they work. (Quoted from *The Economist*, September 12, 1987, p. 83.)

15. If it is assumed that shareholders may be indifferent toward holding shares or having them nationalized at the going market rate, it opens a financing problem for the government that proposes to buy them out.

16. For a few individual companies, the "scorecard" is as follows:

Table 6.4

Estimated Number of Shareholders (in millions)

	Date of privatization	At offer	In October 1987	Implied attrition of shareholders (% p.a.)
British Telecom	November 1984	2.3	1.6	12.0
British Gas	December 1986	4.5	3.0	38.5
British Airways	February 1987	1.0	0.4	74.7
Amersham	February 1982			33.8

Note: Data from *The Economist*, October 24, 1987, p. 29, except for Amersham, *The Economist*, September 13, 1987, p. 84.

17. An illustration of this development—lamented already in the early years of this century by Pareto—is provided by English legislation. Between the Master and Servant Code of 1823, in which breaches of contract by workers were punishable by jail, and the Trades Disputes Act of 1906, which transferred one part of the state prerogatives to unions (union funds were exempt from liabilities resulting from torts) and condoned nonviolent picketing, there are only 83 years but a whole world of difference as concerns the idea of the proper role of the state in industrial legislation. Incidentally, in 1982, the Thatcher government substantially curtailed rights enjoyed by the unions under the 1906 Act.

18. The same holds if the state is called upon to uphold workers' interests (or the interests of any other social group). It should be noted, however, that the workers' movement never adhered to the theory of a neutral state. Its attempt to use the state apparatus for its own purposes is therefore consistent with its professed ideology.

19. In such a case private investors' demand for loss-making state firms may also be forthcoming: if, inclusive of externalities, a firm is efficient, there is such a change in property rights that would make the firm financially attractive for private investors.

20. Yet *The Economist* (October 3, 1987, p. 30) rightly pointed out: "Current proponents of people's capitalism . . . too often confuse the aim of wider share ownership with one means of achieving it: cut-price privatisation, or advertising expensively to persuade not-so-dumb folk . . . to buy for £200 what the newspapers rightly tell them could be sold for £300 tomorrow. This confusion does no good at all to causes of privatisation and popular capitalism."

21. The use of the state for private gain also occurs in nationalizations at above-market price. However, the effects on liberalization are clear: it is an anti-liberalization move both because it increases the state sector and because it gives a subsidy to those from whom the firm is bought.

22. And entrepreneurs.

23. If public (or state) sector firms have strong links with different parties, as in Italy, privatization may be opposed by politicians in general who would thus lose the power to influence decisions of the firms or to channel the money to their favorite areas or pet projects.

24. Ralf Dahrendorf (*Class and Class Conflict in Industrial Society*, Stanford, 1959, p. 301), for example, writes: ". . . the ruling political class of post-capitalist society consists of the administrative staff of the state, the governmental elites at its head and those interested parties which are represented by the governmental elite. This insistence on governmental elites as the core of the ruling class must be truly shocking to anybody thinking in Marxian terms or, more generally, in terms of the traditional concept of class." (Cited in Bell, 1976, p. 52.)

25. The problem is apparent in a conversation between editors of the Polish magazine *Polytika* and the "Solidarity" leadership. The editors question whether trade unions can play the same role in Poland as in the West: "Are the trade unions to represent workers' interests only, since this is their primary role? In the West, it is different. On the one hand, there are trade unions, on the other, owners, and the state plays the role of arbiter. If trade unions make exaggerated claims, the owner will go bankrupt. How would this function here [in Poland]? Since the firm is not the private property of the director, he can yield to, and accept, workers' demands. This has already happened. Would other workers accept subsidizing that firm?" (Quoted from Lay, ed., 1985, p. 244 [my translation—B.M.].)

26. And if they still do not want to become laborers, they would *ipso facto* condemn themselves to disappearance. Their total incomes would sink and, like aristocrats who refused to sully their hands with commerce, they would become queer "marginals."

27. This conflict between workers and owners within the same firm is considered

by Aron (1964, pp. 106–107) to be an inseparable trait of a capitalist society: ". . . dans le cadre de l'entreprise, il y a tension ou lutte entre salariés et propriétaires, chacun songeant à son revenu propre, mais non pas antagonisme irréductible."

28. "Aussi longtemps que subsistent des capitalistes individuels, l'ouvrier ne peut pas ne pas avoir le sentiment qu'il est victime de l'exploitation. L'avantage minimum que vous devez reconnaître à la propriété collective, c'est de supprimer un facteur *psychologique* de conflit." (Aron, 1964, pp. 133–156 [emphasis added—B.M.].)

Notes to Chapter 7

1. James Burnham (1960, p. 178).

2. A change in the legal framework requires revolutionary, and not pragmatic, leaders.

3. In effect, the Soviet regime was the first *modern* totalitarianism in that it primarily used an economic discourse to present its advantages.

4. As Afrikaners seem to be realizing.

5. Prior to that, the language of politics had not yet been conquered by economics (except perhaps in a few of the most developed countries, in particular, Britain and the United States). Economics was not perceived, or thought by the public, to be as important as other, more narrowly political issues like national power, territorial aggrandizement, political stability, or ethnic and religious homogeneity.

6. It is significant that writers as different in their ideological and moral outlooks as Schumpeter, Ellul, and Horvat—but all writing in the 1940s and 1950s—would believe in the relative superiority of state interventionism and centralization. Horvat (1964, p. 93): "The idea that in our century the state plays, or should play, a decisive role in transforming stationary economies into growing economies seems to command a uniquely large acceptance." Schumpeter (1947, pp. 195–196): ". . . socialization [of the instruments of production] means a stride beyond big business on the way that has been chalked out by it or, what amounts to the same thing, that socialist management may conceivably prove superior to big business capitalism as big business capitalism has proved to be to the kind of competitive capitalism of which the English industry of a hundred years ago was the prototype." Ellul (1964, p. 184): "It would seem that we are today unable to escape the facts. And the facts direct us toward the planned economy, regardless of our theoretical judgments in the matter."

7. Although we have proposed (see chapter 3) some arguments why a liberal regime may generally be more efficient economically than a state-oriented one, confirmation of that proposition is not indispensable to the pragmatic argument that liberalization now would yield better results than increased regulation.

8. This argument is developed at greater length in section 4.2.

9. Note that the optimistic view concerning the efficiency of large corporations was fashioned after the experience of World War II and the corporations' economic performance at the time. A parallel effect of the two world wars on economic thinking is obvious.

10. Yet at the outbreak of World War I, the Serbian deputy head of the General Staff was in a spa in Austria. He was allowed to return without any obstacles and take over the command of the Serbian army, which then proceeded to fight the Austrians.

11. Observing resemblances in customs and social structure in Germany, France and England in the fourteenth century, Tocqueville (1955, p. 15) writes: "The community was divided up on the same lines and there were the same hierarchy of classes. The nobles held identical positions, had the same privileges, the same appearance; there was, in fact, a family likeness between them, and one might almost say they were not different men but essentially the same men everywhere."

The same idea is expressed by Mises (1979, pp. 24–25): "... in those ages—in which the status societies were predominant in Europe, as well as in the colonies which Europeans later founded in America—people did not consider themselves to be connected in any special way with the other classes of their own nation; they felt much more at one with the members of their own class in other countries. ... The most visible effect of this state of affairs was the fact that the aristocrats all over Europe used the same language. And this language was French, a language not understood, outside France, by other groups of the population."

12. The alarm created by the French Revolution is well described by Tocqueville (1955, pp. 3–4): "What . . . had seemed to European monarchs and statesmen a mere passing phase . . . was now discovered to be something absolutely new, quite unlike any previous movement, and so widespread, monstrous, and incalculable as to baffle human understanding."

13. The Revolutionary War, as Salvemini writes (1962, p. 261), "... appeared to them [revolutionaries] as different in character from any that occurred in the past. It was not a conflict between two monarchs, whose armies obediently killed one another in the course of some territorial dispute. It was the struggle of the whole people against the armies of several kings, not in defence of a city or frontier province, but to safeguard their own independence; a struggle which they urged the opposing peoples to join, affirming that France had taken up arms not against them but against their rulers; offering to help them in gaining their own freedom, and opening the doors of their country to those who deserted the flag of despotism and rallied to the side of freedom."

14. It may be recalled that one of the reasons why the German General Staff allowed Lenin and his group to move from Switzerland to Russia was exactly because of the Bolsheviks' extremism. They were thought good enough to provoke trouble for the government and thus help the Germans, but not likely to take power.

15. In this context, it is interesting to contrast Stalin's role to that of Napoleon. Stalin was often accused not only of betraying the Revolution, but of bringing about its Thermidor. This clearly is not true. Although Stalin, in a revision of Marxism, opted for "socialism in one country" he did not heal the split between the two elites, but rather solidified it, by creating in the Soviet Union a new elite fundamentally different (in its social status, origin, and claim to power) from the one in capitalism.

16. On the split of the elites Simon Kuznets writes (1977, pp. 12–13): "In the less developed countries, like Russia and China, the heretofore gradually growing modern elements were weakened by World Wars I and II, sufficiently to give way to Communism. Among the developed countries of Europe, the first World War led to the dissolution of a multinational monarchy like Austria-Hungary, and the emergence of fascism in Italy, Germany, and some of the other European states—another case of the use of a hierarchically organized dictatorial party to force the growth of economic and political power of the country by ideologically claimed control over the population. Since these were new approaches, representing violent breaks with the past, explicit hostility to the past, and to other nations still associated with it . . . became a long-term policy at times taking a particularly virulent form. These outbreaks of deviant and self-proclaimed revolutionary regimes . . . introduced into the world, particularly after the 1920s, elements of cleavage and divisiveness that were absent, or only latent, before World War I."

17. These examples show that major revolutions, like the Christian, the French, and the Russian (in the Western World), which are formulated in a universalist language, are by necessity expansionist and cause a split in the world elite, simply because they bring to power an entirely different class of people. The same applies today to the Iranian revolution: it is radical, universalist, and proselytizing.

18. Can the French agricultural policy be considered in isolation from the rest of

the EEC, or even from Australia's farm policy? Or could the U.S. budget deficit be looked at in abstraction from the West German interest rates? It may be conjectured that the next step toward the unification of Western economic policies would be the creation of a common currency. The existence of a common European unit of account (ECU) may be the first step in that direction.

19. In other words, even if one were to concentrate on purely physical criteria in order to establish the efficiency of the firm (i.e., on the fulfillment of plan targets) distortions over which the firm has no control, as for example, nondelivery of required inputs, make even such an assessment difficult.

20. Losses could be reduced if there is no decentralization in the decision-making. This, however, is no different from the existing situation in socialist countries, where trade is conducted mostly by the state alone. Yet the purpose of the reforms is a movement toward a decentralized type of coordination among the firms. It is difficult to see decentralization taking place on the domestic market alone, while foreign trade remains the preserve of the state. This would also be inconsistent with another objective of the reform—devolution of the decision-making authority.

In general, decentralization in the foreign sector must always be much more cautious than in the domestic area. A good example of what happens when state ownership of capital is combined with a strong role of politicians and a decentralization in relations with abroad is provided by Yugoslavia. Most of the external debt was contracted by independent firms, which, enjoying a soft budget constraint, knew that ultimately they would not have to pay it off. So they accepted high interest rates. Foreign banks, knowing well that the government would not risk being cut from the international financial markets if the firms failed to pay back, and also that for ideological reasons it would not let the foreign creditors take over the insolvent firms (something the banks were not keen to do anyway), lent generously. Every player in the game behaved optimally . . . and won. The banks are being paid high interest rates, the borrowers cheerfully spent the money, and it is the country as a whole (and in particular those who did not borrow) that is paying the bill. Certainly, if the borrowing policy were either more centralized or the firms entirely independent (with the government willing to give insolvent firms away to foreign creditors), the outcome would have been better.

21. In addition there is a loss of privileges due to increased commercialization and awareness of the monetary costs and benefits that accompany the reform. Previously subsidized vacation spots where the elite stayed at low cost may now become commercial establishments open to anyone. Even if bureaucrats' incomes remain high, they will now have to pay an economic price for a service they are used to receiving practically free. Moreover, people in more successful occupations will now become the most prized guests. And even if the government decided to directly subsidize a resort so that the people in power could continue to enjoy its benefits (undisturbed by the others), greater social awareness of costs will make these subsidies more apparent and thus more difficult to sustain.

22. This also assumes that they hold the liberal conviction that pursuit of wealth does not necessarily presuppose territorial control (i.e., resources can be obtained through international trade without direct political domination). If, as in mercantilism, "wealth and welfare are directly associated with territorial control, force has a high utility [even] within the economic sphere" (Buzan, 1984, p. 603).

Interdependencies created through trade also help peace. As Montesquieu wrote in *Esprit des Lois*: "The natural effect of commerce is to lead to peace. Two nations that trade together become mutually dependent: if one has an interest in buying, the other has one in selling: and all unions are based on mutual needs." (Quoted from Hirschman, 1977, p. 80.)

23. It may also be true that a greater integration of the USSR may not be in the national interest of the United States. Voluntary abdication of a role in the world economy by the USSR has left the U.S. as the only major player in both the economic and political spheres. But this is altogether a different issue which concerns the national interests of a country, not prospects for an advent of the technocratic elites, or for peace.

REFERENCES

Aron, Raymond (1964). *La Lutte de Classes*. Paris: Gallimard.
———— (1965). *Main Currents in Sociological Thought*, vol. 1. London: Pelican.
———— (1967). *Eighteen Lectures on Industrial Societies*. London: Weidenfeld and Nicolson.
Askanas, Benedykt and Kazimierz Laski (1985). "Consumer Prices and Private Consumption in Poland and Austria." *Journal of Comparative Economics*, June, pp. 164–177.
Aubert, Claude (1984). "La Nouvelle Politique Économique dans les Campagnes Chinoises." *Le Courrier des Pays de l'Est*, July-August, pp. 3–32.
Bell, Daniel (1976). *The Coming of Post-Industrial Society*. New York: Basic Books. First published in 1973.
Bergson, Abram (1987). "Comparative Productivity: The USSR, Eastern Europe and the West." *American Economic Review*, June, p. 347.
Blaha, Jaroslav (1986). "Tchecoslovaquie." In *Panorama de l'Europe de l'Est, Le courrier de l'Europe de l'Est*, August-September-October.
Blau, David M. (1987). "A Time Series Analysis of Self-Employment in the United States." *Journal of Political Economy*, No. 3.
Bos, Dieter (1986). *Public Enterprise Economics: Theory and Applications*. Amsterdam; New York: North-Holland.
Braudel, Fernand (1984). *The Perspective of the World*, vol. 3. New York: Harper and Row.
Brittain, Samuel (1986). "Privatization: A Comment on Kay and Thomson." *The Economic Journal*, March, pp. 33–38.
Burnham, James (1960). *The Managerial Revolution*. Bloomington: Indiana University Press. First published in 1941.
Buzan, Barry (1984). "Economic Structure and International Security: the Limits of the Liberal Case." *International Organization*, Autumn, p. 603.
Carr, E.H. (1966). *The Bolshevik Revolution*, vol. 2. London: Penguin Books. First published in 1952.
Cranston, Ross (1987). "Privatisation: A Critique" in Peter Abelson (ed.), *Privatisa-

tion: An Australian Perspective. Sydney: Australian Professional Publications.

Dubravčić, Dinko (1970). "Labor as Entrepreneurial Input: An Essay in the Theory of Producer Cooperative Economy." *Economica*, August, pp. 297–310.

Ellul, Jacques (1964). *The Technological Society.* New York: Vintage Books. First published in 1954.

Friedmann, Wolfgang, ed. (1974). *Public and Private Enterprise in Mixed Economies.* New York: Columbia University Press.

Galbraith, John Kenneth (1967). *The New Industrial State.* London: Penguin Books.

Gams, Andrija (1987). *Svojina.* Belgrade: CFDT.

Gligorov, Vladimir (1984). *Gledišta i sporovi o industrijalizaciji u socijalizmu.* Belgrade: CFDT.

Havlik, Peter (1985). "A Comparison of Purchasing Power Parity and Consumption Levels in Austria and Czechoslovakia." *Journal of Comparative Economics*, June, pp. 178–190.

Hayek, Friedrich A. (1944). *The Road to Serfdom.* Chicago: University of Chicago Press.

———— (1945). "The Use of Knowledge in Society." *American Economic Review*, September.

Hirschman, Albert (1977). *The Passions and the Interests.* Princeton: Princeton University Press.

Hobsbawm, Eric (1979). *The Age of Capital.* New York: Mentor. First published in 1975.

Hoover, Calvin (1959). *The Economy, Liberty and the State.* New York: The Twentieth Century Fund.

Horvat, Branko (1964). *Towards a Theory of Planned Economy.* Belgrade: Yugoslav Institute of Economic Research. First published in 1961.

———— (1982). *Political Economy of Socialism.* Armonk, New York: M.E. Sharpe.

International Monetary Fund (1986). *International Financial Statistics, Yearbook 1986.* Washington, D.C.

Kay, J.A. and D.J. Thomson (1986). "Privatization: A Policy in Search of a Rationale." *The Economic Journal*, March, pp. 18–32.

Kornai, Janos (1986). *Contradictions and Dilemmas.* Cambridge, Massachusetts: MIT Press.

———— (1986a). "The Hungarian Reform Process: Visions, Hopes, and Reality." *Journal of Economic Literature*, December, pp. 1687–1737.

Kuznets, Simon (1977). "Two Centuries of Economic Growth: Reflections on U.S. Experience." Richard T. Ely lecture. *American Economic Review Papers and Proceedings*, February, pp. 1–15.

Lay, Vladimir, ed. (1985). *Društveni pokreti i politički sistem u Poljskoj, 1956–1981: Dokumenti.* Belgrade: Institut drustvenih nauka.

Luttwak, Edward (1983). *The Grand Strategy of the Soviet Union.* New York: St. Martin's Press.

MacDonald, Oliver, ed. (1981). *The Polish August: Documents from the Beginnings of the Polish Workers' Rebellion.* San Francisco: Left Bank Books.

Mises, Ludwig von (1963). *Human Action*, revised edition. Chicago: Henry Regnery Company. First published in 1949.

———— (1979). *Economic Policy.* Chicago: Regnery/Gateway.

Nuti, Mario D. (1987). "Feasible Financial Innovation Under Market Socialism" (mimeo). Paper presented on *Workshop on Financial Reform in Socialist Economies*, Florence.

Park, Young C. (1987). "Evaluating the Performance of Korea's Government Invested Enterprises." *Finance and Development*, June.

Perkins, D.H. (1988). "Reforming China's Economic System." *Journal of Economic Literature*, June.

Polanyi, Karl (1957). *The Great Transformation*. Boston: Beacon Press.

Sah, Edward K. (1986). "Major Contradictions in Peking's Economic Reforms." *Issues and Studies: A Journal of China Studies and International Affairs*, December, p. 45.

Salvemini, Gaetano (1962). *The French Revolution, 1788–1792*. New York: Norton. First published in 1905.

Schumpeter, Joseph Alois (1947). *Capitalism, Socialism and Democracy*, 3rd edition. Harper Colophon Books. First published in 1942.

Short, R.P. (1984). "The Role of Public Enterprises: An International Statistical Comparison." In Floyd, Robert H. and Clive S. Gray (eds.), *Public Enterprise in Mixed Economies: Some Macroeconomic Aspects*. Washington: International Monetary Fund.

Stajner, Rikard (1986). "Komparativni pregled nekih elemenata ekonomskog i društvenog razvoja Jugoslavije." *Ekonomski pregled*, No. 5–6, pp. 203–215.

Szul, Roman and Edward Kirejczyk (1987). "Dilemmas of Economic Reform and Self-Management in Poland." *Economic Analysis and Workers' Management*, No. 3, pp. 373–391.

Tao Youzhi (1987). "A Brief Discussion of the Consolidation and Development of the Individual Economy." *Chinese Economic Studies*, Fall, pp. 37–42.

Terrell, Katherine (1987). "Western Capital and Productivity in Polish Industry." University of Pittsburgh, mimeo, July.

Tocqueville, Alexis de (1955). *The Old Regime and the French Revolution*. New York: Doubleday.

Tucker, Robert C., ed. (1978). *The Marx-Engels Reader*, 2nd edition. New York: W. W. Norton and Co.

Vickers, John and George Yarrow (1988). *Privatization: An Economic Analysis*. Cambridge, Massachusetts: MIT Press.

Ward, Benjamin (1958). "The Firm in Illyria: Market Syndicalism." *American Economic Review*, September, pp. 566–589.

Westergaard, John and Henrietta Resler (1975). *Class in a Capitalist Economy*. London: Penguin.

Wicksell, Knut (1970). *Value, Capital and Rent*. New York: Augustus A. Kelley. First published in 1893.

World Bank (1987). *Poland: Reform, Adjustment and Growth*, vol. 1. Country Economic Memorandum, Washington, D.C.

NAME INDEX

Aron, Raymond, 8, 163, 168
Askanas, Benedykt, 160
Aubert, Claude, 162

Bergson, Abram, 159
Blaha, Jaroslav, 163
Blau, David M., 155
Bos, Dieter, 157
Braudel, Fernand, 7, 154
Brittain, Samuel, 110
Burnham, James, 128
Buzan, Barry, 170

Carr, E. H., 158
Clausewitz, 170
Cranston, Ross, 111

Dahrendorf, Ralf, 167
Drago, Roland, 157
Dubravčić, Dinko, 155

Ellul, Jacques, 42, 168

Friedmann, Wolfgang, 156

Galbraith, John K., 133
Gams, Andrija, 153

Havlik, Peter, 160
Hayek, Friedrich A., 16, 39

Hobsbawm, Eric, 145, 164
Hoover, Calvin, 158
Horvat, Branko, 158, 168

Kay, J. A., 116
Keynes, John M., 159
Kirejczik, Edward, 162
Kornai, Janos, 61, 75, 156, 161, 162, 163
Kuznets, Simon, 169

Lange, Oskar, 158
Laski, Kazimierz, 160
Lay, Vladimir, 167
Luttwak, Edward, 158

MacDonald, Oliver, 165
Machiavelli, Niccolo, 37
Marx, Karl, 3, 5, 89, 158, 165
Mises, Ludwig von, 27, 163, 169
Montesquieu, 170
Moore, John, 96

Nuti, Mario, 154

Park, Young C., 154
Perkins, Dwight H., 161
Polanyi, Karl, 35

Rakowski, Miaczyslaw, 55

Resler, Henrietta, 155
Ricardo, David, 157

Sah, Edward K., 162
Salvemini, Gaetano, 169
Schumpeter, Joseph, 132, 153, 157, 159,
    164, 168
Short, R. P., 165
Stajner, Rikard, 160
Szul, Roman, 162

Tao Yuzhi, 162
Terrell, Katherine, 160

Thomson, D. J., 116
Tocqueville, Alexis de, 82, 168, 169
Trotsky, Lev, 127

Vickers, John, 112

Ward, Benjamin, 155
Westergaard, John, 155
Wicksell, Knut, 155

Yarrow, George, 112

Županov, Josip, 161

SUBJECT INDEX

Afghanistan, 147
Albania, 146
Argentina, 163
Asiatic mode of production, 5
Australia, 15, 111
Austria, 14, 90, 160, 168

Babeuf, Gracchus, 134
Bangladesh, 17
Belgium, 19
Bolivia, 17
Botswana, 17
Brecht, Bertolt, 138
Brezhnev, Leonid, 93
Burke, Edmund, 134

Canada, 15
Capitalist economies (system), 14, 22,
    31ff.;
  capitalists and privatization, 113–15;
  central conflict in, 119ff.;
  deregulation, 105–108;
  economic structure of, 96–99;
  future of, 124;
  liberalization in, 103–105, 106;
  political elites, 101, 161;
  privatization, 109ff., 115ff.;
  social cohesiveness, 107–108;
  social composition in, 99ff.;
  state sector evolution, 115–19;

subsidies and protection, 108–109
Capitalist firm
  definition, 6;
  entrepreneurial income in, 6, 39;
  joint-stock company, 6, 96;
  in socialism, 22, 56
Centralization
  and inefficiency, 39, 41, 48
Chile, 17
China, 19, 20, 21, 52, 55–56, 58n, 61–
    62, 65, 69, 72–73, 95, 129, 131,
    137, 145, 161, 163
Communes, 6, 162;
  reform in, 91–93
Communist parties
  attitude toward the state, 46;
  electoral results, 158;
  evolution of, 144–45;
  opposition to reform, 142
Competition
  and laissez-faire, 35–36, 106;
  and monopoly, 35;
  and privatization, 110
Conservative parties
  electoral results, 43
Cooperative–2
  definition, 11
Coordination, economic, 11;
  centralized-decentralized, 12, 33
Cuba, 16, 138

Czechoslovakia, 19, 20, 58n, 65, 147, 160, 163

Deng Xiaoping, 51, 91, 128
Denmark, 15
Dirigisme, 45, 131–32
Distribution of income, 11;
  in capitalism, 99–101, 104;
  in socialism, 56–57
Disutility of labor
  as cost of production, 157

Eastern Europe
  economic failure, 49ff.
Efficiency
  in the use of capital, 37–38;
  in the use of labor, 40;
  of different modes of production, 39–41
Elites
  segmented, 134–35;
  unified, 133–34, 137
Entrepreneurial firm, 18, 98;
  definition, 7;
  venture capital, 155
Entrepreneurship, 9–10;
  definition, 4;
  firms' maximands and, 13;
  workers as entrepreneurs, 10–11
Ethiopia, 16, 156
Europe, 128;
  integration of Western Europe, 136;
  pre-World War II, 129
Externalities, 38

Feudalism, 5
Finland, 15
France, 15, 19, 43, 45, 88, 131
French revolution, 134–35, 165

Gandhi, Rajiv, 128
Germany, 90, 128, 136, 169;
  East, 19, 20, 161;
  West, 15, 19, 43, 166
Gorbachev, Mikhail, 51, 93, 128, 136
Greece, 15, 160
Grenada, 147
Guatemala, 17
Guinea, 17
Guyana, 16, 17

Hungary, 11, 19, 20, 21, 61, 69, 72–73, 90, 138, 145, 160, 161;

reform in, 55–56
Hu Yaobang, 92

India, 17, 129, 131
Italy, 15, 18, 19, 43, 167
Ivory Coast, 17

Japan, 155, 110, 128, 147
Jaruzelski, Wojciech, 145

Kampuchea, 77
Kenya, 17
Kuwait, 152

Labor-managed firm, 11;
  in capitalism, 19;
  in socialism, 21, 56, 156
Labor-management, 63, 156
Legal framework
  definition, 25, 28
Lenin, Vladimir, 90, 169
Liberalization, economic
  in capitalism, 103–105;
  and competition, 106;
  definition, 34ff.;
  and monopoly, 106;
  and privatization, 110, 115;
  in socialism, 63–64, 65ff;
  trend toward, 42, 131–32
Liberia, 17

Management buy-outs, 6
Mao Zedong, 91, 94, 95n
Market economy
  definition, 34
Mode of production, 22;
  definition, 4;
  efficiency of different modes, 39–41
Mongolia, 16

Napoleon, 135, 169
Nepal, 17
Netherlands, 15, 19, 135
New Zealand, 15
Niger, 17
Norway, 15

Ownership of capital
  economic, 4
Owner-worker firms
  Cooperative-1, 10, 18–19
  definition, 10, 39

ESOPs, 150;
  in capitalism, 18, 97–98;
  in socialism, 21, 56

Pakistan, 17
Paraguay, 17
Philippines, 17, 115
Planned economy
  definition, 34;
  efficiency of, 48, 52n, 158;
  and monopoly, 36;
  prices in, 13n
Poland, 18, 20, 21, 23–24, 26, 50, 55–
    56, 65, 74, 84, 87, 88–91, 145, 160,
    161, 162, 167;
  reform in, 61–3
  "Solidarity," 88–91, 164–65
Portugal, 15
Prices
  equilibrium, 13;
  regulation of, 28
Public corporation, 14;
  definition, 7;
  entrepreneurship in, 9–10;
  in less developed countries, 16–17;
  in OECD countries, 15, 97;
  in socialism, 56
Public sector, 14;
  definition, 8

Reagan, Ronald, 42, 128
Robespierre, Augustin, 134, 135
Romania, 50, 146, 159
Russia, 136, 145–46

Saint-Just, L.A., 134
Sierra Leone, 17
Singapore, 17
Slavery, 5;
  and efficiency, 40
Socialist economies (system), 19ff., 22,
    31ff.;
  agriculture in, 65–7;
  bonds and equities, 69–71;
  central conflict in, 74ff., 142;
  dismantling state sector, 59ff.;
  economic structure of, 55–56;
  evolution, 33–34;
  expanding private sector, 65ff.;
  future of, 145–46;
  high-level bureaucracy, 58, 80;
  liberalization, 71–72;

losses in foreign trade, 141, 170;
middle-level bureaucracy, 58, 75–79;
obstacles to growth, 137;
obstacles to international integration,
    139–44;
political economy of reform, 79ff.;
private sector in, 65, 67–69, 83–85;
social composition of, 57–59;
technocrats in, 75
Socialist parties
  attitude toward the state, 44–45
South Korea, 17, 138, 145, 154
Soviet Union, 51–52, 58n, 65, 131, 137,
    147, 154, 161, 162, 168;
  gulag, 6;
  kolkhozes, 6, 20, 31
  October Revolution, 133, 135, 169,
    171
  reform in, 93–95
Spain, 15, 19, 145
Sri Lanka, 17
Stalin, Joseph, 94, 95n, 165, 169
State interference, 30ff., 106;
  definition, 27;
  discretionary-uniform, 29;
  inefficiency of, 41;
  subsidies and protection, 29;
  taxation, 29, 38
State ownership of capital
  and inefficiency, 38, 41
State sector, 14;
  definition, 8
State-socialist firm
  definition, 7;
  diffused ownership of capital, 37–38;
  information problem, 39;
  in less developed countries, 16–17;
  in OECD countries, 15, 97;
  in socialist countries, 19
Structure of production, 22, 31;
  in China and Hungary, 72–73;
  in Czechoslovakia, 163;
  definition, 14ff.;
  in UK and Poland, 23–26
Sudan, 17
Sweden, 15, 19

Taiwan, 17
Tanzania, 17
Technocrats
  advent of, 128–29, 136;
  and prospects for peace, 146–48;

in socialism, 57, 75–77
Thailand, 17
Thatcher, Margaret, 42, 128
Togo, 17
Trotsky, Lev, 95n
Tunisia, 17
Turkey, 15

United Kingdom, 15, 18, 19, 23ff., 42, 43, 101, 131, 135, 157, 166, 168; privatization in, 110–13

United States, 15, 18, 42, 128, 135, 147, 168, 171

Venezuela, 16, 17
Venice, 7, 169
Vietnam, 16, 146, 147

Yugoslavia, 11, 65, 84, 156, 161, 170; Serbia, 168

Zambia, 16

ABOUT THE AUTHOR

Branko Milanović is an economist with the World Bank in Washington, D.C., where he has done work on Turkey and Poland. He previously worked at the Institute of Economics in Belgrade (Yugoslavia). Milanović earned his Ph.D. in economics from the University of Belgrade.